Probabilistic Topic Models

Di Jiang • Chen Zhang • Yuanfeng Song

Probabilistic Topic Models

Foundation and Application

 Springer

Di Jiang
AI
WeBank
Shenzhen, Guangdong, China

Yuanfeng Song
AI
WeBank
Shenzhen, Guangdong, China

Chen Zhang
Department of Computing and School of
Hotel and Tourism Management
The Hong Kong Polytechnic University
Hong Kong, Hong Kong

ISBN 978-981-99-2433-2   ISBN 978-981-99-2431-8   (eBook)
https://doi.org/10.1007/978-981-99-2431-8

This Springer imprint is published by the registered company Springer Nature Singapore Pte Ltd.
The registered company address is: 152 Beach Road, #21-01/04 Gateway East, Singapore 189721,
Singapore

Preface

As important tools for text mining and analytics, probabilistic topic models receive substantial attention from both academia and industry. Each year a large number of research papers on topic models are published at conferences and in journals. At the same time, topic models are widely applied in the industry and have become important tools for semantic analysis in engineering applications.

Although the field has seen significant progress over the last two decades, the existing literature consists primarily of research papers and technical reports, which do not necessarily follow the same terminologies or mathematical symbols. Furthermore, the existing work typically focuses on designing new models, and those about applying topic models in industrial scenarios are relatively scarce. As the above two obstacles impede the development of this field, a monograph is urgently needed to systematically introduce the technical foundation of probabilistic topic models and their application in the real world.

By unifying the terminologies or mathematical symbols, this book provides readers with a convenient approach to understand the technical details of topic models. Beyond covering the recent advances of academic research, this book provides a panoramic view of how topic models are applied in real-world scenarios. Readers will be instructed on the means to select, design, and apply topic models for specific tasks. We hope this book can bridge the gap between academic research and industrial application, and help topic models to play a more significant role in academia and industry.

This book can be a reference for senior undergraduate, graduate students, and software engineers. Given the broad application of topic models in the industry, the book can serve as a reference for job seekers preparing for interviews.

This book is divided into five parts. The first part (Chap. 1) introduces the fundamental principles of linear algebra, probability theory, and Bayesian networks. The second part (Chap. 2) introduces the topic models used in real-world scenarios. The third part (Chaps. 3 to 9) introduces the methods used to create high-quality topic models. The fourth part (Chap. 10) introduces practical applications of topic models in real-world scenarios. Finally, this book presents a catalogue of the most important topic models from the literature over the past decades, which can be referenced and indexed by researchers and engineers in related fields.

The views expressed in this book are those of the authors and do not necessarily reflect the opinions or positions of the institution to which the authors belong. After the first draft, many friends helped to review some chapters and put forward valuable suggestions. The authors would like to express many heartfelt thanks to them.

Shenzhen, China   Di Jiang
Hong Kong, Hong Kong   Chen Zhang
Shenzhen, China   Yuanfeng Song
Feb, 2022

Contents

Chapter 1
Basics

This chapter introduces the basic concepts of linear algebra, probability theory and Bayesian networks. Readers familiar with these concepts may skip this chapter and proceed to Chap. 2.

1.1 Linear Algebra

Linear algebra deals with concepts like vectors, matrices, linear transformations, and linear equations.

1.1.1 Vector

A vector is a mathematical object with a magnitude and a direction. The number of elements in a vector is known as the vector's dimension. Vectors are typically denoted using bold and lowercase letters such as x. Each element of vectors can be defined explicitly by writing $x = (x_1, x_2, \ldots, x_n)$. Vectors can be expressed vertically or horizontally, known as row vectors or column vectors, respectively. Figure 1.1 shows an m-dimension column vector and an n-dimension row vector.

1.1.2 Matrix

A matrix is a rectangular arrangement of $m \times n$ elements, composed of m rows and n columns. Matrices are typically denoted using bold and uppercase letters, such

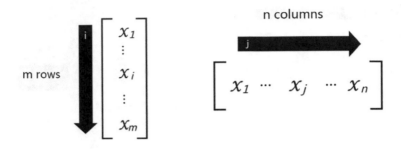

(a) m-dimension column vector   **(b)** n-dimension row vector

Fig. 1.1 Examples of a row vector and a column vector

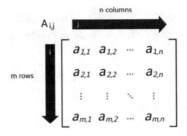

Fig. 1.2 A matrix with m rows and n columns

as A. Figure 1.2 shows an example of $m \times n$ matrix. The matrix element of row i and column j is denoted as $A_{i,j}$. It is apparent that a vector is a special case of a matrix: an $m \times 1$ matrix is an m-dimensional column vector and a $1 \times n$ matrix is an n-dimensional row vector. A matrix that has the same number of rows and columns is known as a square matrix. A matrix in which every element along the main diagonal has the value 1 and every other element has the value 0 is known as an identity matrix. An identity matrix of order n is typically denoted as I_n.

An example of a 2×3 matrix:

$$A = \begin{bmatrix} 1 & 2 & 3 \\ 2 & 1 & 5 \end{bmatrix}$$

An example of a 2×2 matrix:

$$A = \begin{bmatrix} 1 & 2 \\ 2 & 1 \end{bmatrix}$$

An example of an identity matrix:

$$A = \begin{bmatrix} 1 & 0 \\ 0 & 1 \end{bmatrix}$$

1.1.3 Matrix Operations

We proceed to introduce common matrix operations: addition, subtraction, multiplication, transposition, and inversion.

1.1.3.1 Matrix Addition and Subtraction

Addition and subtraction are defined for two matrices that have the same number of rows and columns. These operations are performed by adding or subtracting each element in one matrix from the corresponding element with the same row and column in the other matrix. Consider an $m \times n$ matrix $\mathbf{A}$ and an $m \times n$ matrix $\mathbf{B}$. Each element in matrix $\mathbf{C}$ ($\mathbf{C} = \mathbf{A} + \mathbf{B}$) is calculated as follows:

$$C_{i,j} = A_{i,j} + B_{i,j} \tag{1.1}$$

where $1 \leq i \leq m$, $1 \leq j \leq n$. Similarly, each element in matrix $\mathbf{D}$ ($\mathbf{D} = \mathbf{A} - \mathbf{B}$) is calculated as follows:

$$D_{i,j} = A_{i,j} - B_{i,j} \tag{1.2}$$

We present two examples as follows:

$$\begin{bmatrix} 1 & 3 \\ 4 & 2 \\ 4 & 6 \end{bmatrix} + \begin{bmatrix} 6 & 8 \\ 9 & 1 \\ 5 & 2 \end{bmatrix} = \begin{bmatrix} 7 & 11 \\ 13 & 3 \\ 9 & 8 \end{bmatrix}$$

$$\begin{bmatrix} 1 & 3 \\ 4 & 2 \\ 4 & 6 \end{bmatrix} - \begin{bmatrix} 6 & 8 \\ 9 & 1 \\ 5 & 2 \end{bmatrix} = \begin{bmatrix} -5 & -5 \\ -5 & 1 \\ -1 & 4 \end{bmatrix}$$

1.1.3.2 Matrix Multiplication

Multiplication is defined for matrices $\mathbf{A}$ and $\mathbf{B}$ where the number of columns in $\mathbf{A}$ is equal to the number of rows in $\mathbf{B}$. The product of an $m \times n$ matrix $\mathbf{A}$ and an $n \times p$

matrix $\mathbf{B}$ is an $m \times p$ matrix $\mathbf{C}$, i.e., $\mathbf{C} = \mathbf{AB}$. Each element of $\mathbf{C}$ is calculated as follows:

$$AB_{i,j} = A_{i,1}B_{1,j} + A_{i,2}B_{2,j} + \ldots + A_{i,n}B_{n,j} = \sum_{r=1}^{n} A_{i,r}B_{r,j} \qquad (1.3)$$

where $1 \le i \le m$, $1 \le j \le p$. An example is given as follows:

$$\begin{bmatrix} 1 & 4 \\ 2 & 3 \end{bmatrix} \times \begin{bmatrix} 2 & 4 & 5 & 8 \\ 6 & 3 & 4 & 5 \end{bmatrix} = \begin{bmatrix} 26 & 16 & 21 & 28 \\ 22 & 17 & 22 & 31 \end{bmatrix}$$

1.1.3.3   Matrix Transposition

Matrix transposition describes the process by which each matrix element at row j and column i is exchanged with the element at row i and column j. The transposition of an $m \times n$ matrix $\mathbf{A}$ produces an $n \times m$ matrix $\mathbf{A}^{\mathrm{T}}$. Each element of $\mathbf{A}^{\mathrm{T}}$ is calculated as follows:

$$(A^{\mathrm{T}})_{i,j} = A_{j,i} \qquad (1.4)$$

An example is given as follows:

$$\begin{bmatrix} 1 & 2 \\ 2 & 3 \\ 5 & 1 \end{bmatrix}^{\mathrm{T}} = \begin{bmatrix} 1 & 2 & 5 \\ 2 & 3 & 1 \end{bmatrix}$$

1.1.3.4   Matrix Inversion

Consider two square $n \times n$ matrices $\mathbf{A}$ and $\mathbf{B}$ that satisfy:

$$\mathbf{AB} = \mathbf{BA} = \mathbf{I}_n \qquad (1.5)$$

where $\mathbf{I}_n$ is an identity matrix. The matrix $\mathbf{B}$ is referred to as the inverse matrix of the matrix $\mathbf{A}$ and can be denoted as $\mathbf{A}^{-1}$. The matrix $\mathbf{A}$ is referred to as an invertible matrix. It is obvious that an invertible matrix must be a square matrix.
An example is given as follows::

$$\mathbf{A} = \begin{bmatrix} 1 & 3 \\ 2 & 4 \end{bmatrix}, \mathbf{A}^{-1} = \begin{bmatrix} -2 & 1.5 \\ 1 & -0.5 \end{bmatrix}$$

1.1.4   Orthogonal Matrix

A square matrix $\mathbf{A}$ is known as an orthogonal matrix if its elements are real and it satisfies $\mathbf{A}^T = \mathbf{A}^{-1}$. For example,

$$\mathbf{A} = \begin{bmatrix} \frac{1}{3} & \frac{2}{3} & \frac{2}{3} \\ \frac{2}{3} & \frac{1}{3} & -\frac{2}{3} \\ \frac{2}{3} & -\frac{2}{3} & \frac{1}{3} \end{bmatrix}$$

is an orthogonal matrix, because:

$$\mathbf{A}^T\mathbf{A} = \begin{bmatrix} \frac{1}{3} & \frac{2}{3} & \frac{2}{3} \\ \frac{2}{3} & \frac{1}{3} & -\frac{2}{3} \\ \frac{2}{3} & -\frac{2}{3} & \frac{1}{3} \end{bmatrix} \begin{bmatrix} \frac{1}{3} & \frac{2}{3} & \frac{2}{3} \\ \frac{2}{3} & \frac{1}{3} & -\frac{2}{3} \\ \frac{2}{3} & -\frac{2}{3} & \frac{1}{3} \end{bmatrix} = \begin{bmatrix} 1 & 0 & 0 \\ 0 & 1 & 0 \\ 0 & 0 & 1 \end{bmatrix}$$

1.1.5   Eigenvalues and Eigenvectors

Given an $n \times n$ square matrix $\mathbf{A}$, suppose we have an n-dimensional and nonzero column vector $\mathbf{v}$ and a scalar λ that satisfy

$$\mathbf{A}\mathbf{v} = \lambda\mathbf{v} \tag{1.6}$$

the scalar λ is referred to as an eigenvalue of $\mathbf{A}$; the vector $\mathbf{v}$ is referred to as an eigenvector of $\mathbf{A}$ corresponding to the eigenvalue λ. The above equation can also be rewritten as:

$$(\mathbf{A} - \lambda\mathbf{I})\mathbf{v} = 0 \tag{1.7}$$

Eigenvalues and eigenvectors have certain special properties:

- Given a set of all eigenvalues $\lambda_1, \lambda_2,\ldots,\lambda_n$ of an n-order square matrix $\mathbf{A}$, we have

$$\lambda_1 + \lambda_2 + \ldots + \lambda_n = \sum_{i=1}^{n} a_{i,i} \tag{1.8}$$

$$\lambda_1\lambda_2\ldots\lambda_n = |A| \tag{1.9}$$

- Suppose we have the eigenvalue λ and the corresponding eigenvector $\mathbf{v}$ of an invertible matrix $\mathbf{A}$, the value $\frac{1}{\lambda}$ is also an eigenvalue with the corresponding eigenvector $\mathbf{v}$.

- Given the eigenvalue λ and the corresponding eigenvector $\mathbf{v}$ of a square matrix $\mathbf{A}$, the value λ^m is an eigenvalue of $\mathbf{A}^m$ with the corresponding eigenvector $\mathbf{v}$.

For example, the eigenvalues and eigenvectors of the following matrix $\mathbf{A}$ is as follows:

$$\mathbf{A} = \begin{bmatrix} -1 & 1 & 0 \\ -4 & 3 & 0 \\ 1 & 0 & 2 \end{bmatrix}$$

The characteristic polynomial of $\mathbf{A}$ is:

$$|\mathbf{A} - \lambda \mathbf{I}| = \begin{vmatrix} -1-\lambda & 1 & 0 \\ -4 & 3-\lambda & 0 \\ 1 & 0 & 2-\lambda \end{vmatrix} = (2-\lambda)(1-\lambda)^2 \tag{1.10}$$

Therefore, the eigenvalues of $\mathbf{A}$ are: $\lambda_1 = 2$ and $\lambda_2 = \lambda_3 = 1$.

As an example, for λ_1, the corresponding eigenvector $\mathbf{v}_1$ should satisfy the equation $(\mathbf{A} - 2\mathbf{I})\mathbf{v}_1 = 0$, that is

$$(\mathbf{A} - 2\mathbf{I})\mathbf{v}_1 = \begin{pmatrix} -3 & 1 & 0 \\ -4 & 1 & 0 \\ 1 & 0 & 0 \end{pmatrix} \begin{pmatrix} x_1 \\ x_2 \\ x_3 \end{pmatrix} = 0 \tag{1.11}$$

The eigenvector $\mathbf{v}_1$ is:

$$\mathbf{v}_1 = \begin{pmatrix} 0 \\ 0 \\ 1 \end{pmatrix}$$

1.2 Probability Theory

This section introduces probability concepts that are relevant for the study of topic models.

1.2.1 Probability Distribution

In probability theory, a random event refers to a process whose outcomes cannot be predicted. The real world is full of various random events, such as the number of visitors to a store during a certain period of time and the time at which a bus arrives at a stop. To analyze the random events, the concepts of random variables are proposed as a measurement. Such variables are typically represented by lowercase letters, such as x and y, and can be either discrete or continuous. For example, the number of visitors to a store during a given time period can only take integer values. Whereas, for a bus route scheduled every 20 minutes, the time at which a bus arrives at a stop is a continuous variable within the range $[0, 20)$. Probability theory quantifies the possibility of random events occurring. The probability that the random variable x obtains a specific value a is denoted by $p(x = a)$, often abbreviated as $p(a)$. Probability has the following basic properties:

- For any event a, $0 \leq p(a) \leq 1$.
- The probability of an inevitable event a is 1, that is, $p(a) = 1$.
- The probability of an impossible event a is 0, that is, $p(a) = 0$.

A probability distribution describes all possible values that a random variable can take and the probability that the variable takes each value. A probability density function returns the probability that a continuous random variable takes a given input value. Figure 1.3 shows a uniform probability distribution in the interval $[a, b]$, for which the corresponding probability density function is:

$$p(x) = \begin{cases} \frac{1}{b-a} & a \leq x \leq b \\ 0 & \text{otherwise} \end{cases} \tag{1.12}$$

A joint probability distribution is the probability distribution of two or more random variables. For example, consider flipping a coin twice, and using x and y to represent the results of the first and the second flips. The random variable takes on value 1 if the coin lands on heads, or 0 if the coin lands on tails. Then, $p(x = 1, y - 0)$ represents the probability that the first coin lands on heads and the second coin lands on tails. For discrete random variables, the joint probability distribution can be presented in the form of a table or a function; for continuous

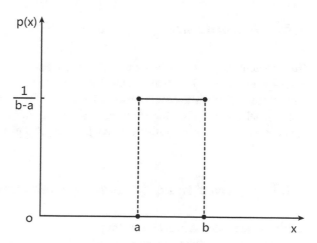

Fig. 1.3 A uniform probability distribution in the interval $[a, b]$

random variables, the joint probability distribution is expressed by the integral of the joint probability density function.

A marginal distribution is the probability distribution of a subset of variables from a group of random variables. For example, the marginal distribution of the random variable x, which is a part of the joint probability distribution $p(x, y)$ of two random variables, can be calculated as:

$$p(x) = \sum_y p(x, y) \tag{1.13}$$

A conditional probability distribution is the probability distribution of an event subject to the occurrence of another event. For example, given that the first coin flip is tail (i.e., $x = 0$), the probability that the second coin flip returns head (i.e., $y = 1$) is denoted as $p(y = 1|x = 0)$. The conditional probability can be calculated as follows:

$$p(x = a|y = b) = \frac{p(x = a, y = b)}{p(y = b)} \tag{1.14}$$

The chain rule of probability theory is primarily used to calculate the joint probability distribution of multiple random variables. The chain rule is mathematically defined as follows:

$$p(x_1, \ldots, x_n) = p(x_1) \prod_{i=2}^{n} p(x_i \mid x_1, \ldots, x_{i-1}) \tag{1.15}$$

As an example, chain rule can be used to calculate the joint probability $p(x, y, z)$ as follows: $p(x, y, z) = p(x|y, z)p(y, z) = p(x|y, z)p(y|z)p(z)$.

1.2.2 Independence

Two random variables are independent of each other if the distribution of one does not depend on the other. Mathematically, if $p(x, y) = p(x)p(y)$, then x and y are independent. Furthermore, suppose that we have three variables x, y and z, if the conditional distribution of x given z does not depend on the value of y, then the random variables x is conditionally independent of y given z.

1.2.3 Expected Value, Variance, and Standard Deviation

The expected value reflects the average value that a random variable is expected to take in random experiments, and it is typically denoted as μ or $E(x)$. Let us assume that g is a continuous function and $Y = g(X)$. For a discrete random variable X, $p(x)$ represents the probability that X obtains the value of x and the expected value of Y is:

$$\mu = E(Y) = E[g(X)] = \sum_x g(x)p(x) \qquad (1.16)$$

For a continuous random variable X, $f(x)$ represents the probability density function and the expected value of Y is:

$$\mu = E(Y) = E[g(X)] = \int g(x)f(x)\mathrm{d}x \qquad (1.17)$$

Variance and standard deviation are used to describe the difference between the observed values of a variable and its expected value. The two quantities describe the degree of dispersion of a random variable. The greater the variance and the standard deviation, the higher the degree of dispersion. Variance is typically denoted by σ^2; standard deviation is typically denoted by σ. These values can be calculated as follows:

$$\sigma^2 = \frac{\sum (x - \mu)^2}{N} \qquad (1.18)$$

$$\sigma = \sqrt{\frac{\sum (x - \mu)^2}{N}} \qquad (1.19)$$

where N is the total number of samples.

1.2.4 Common Probability Distributions

In this section, we introduce some probability distributions used in the context of topic models, such as the gamma distribution, the binomial distribution, the multinomial distribution, the beta distribution, and the Dirichlet distribution.

The gamma distribution is a continuous probability function that is defined by two parameters: the shape parameter α and the scale parameter β. The probability density function of the gamma distribution is defined as follows:

$$p(x|\alpha, \beta) = \frac{x^{(\alpha-1)} \lambda^\alpha e^{-\lambda x}}{\Gamma(\alpha)}, x > 0 \qquad (1.20)$$

The function $\Gamma(\cdot)$ is known as the gamma function, which is defined as:

$$\Gamma(z) = \int_0^\infty \frac{t^{z-1}}{e^t} dt \tag{1.21}$$

The expected value and variance of the gamma distribution are:

$$E(x) = \frac{\alpha}{\beta} \tag{1.22}$$

$$Var(x) = \frac{\alpha}{\beta^2} \tag{1.23}$$

The Bernoulli distribution is a discrete probability distribution. It has a single parameter $\theta(0 < \theta < 1)$, and the probability of taking the values 1 and 0 is θ and $1 - \theta$, respectively. Its probability density function is:

$$p(x|\theta) = \theta^x (1 - \theta)^{1-x} \tag{1.24}$$

where x takes the value 0 or 1. The expected value and variance of the Bernoulli distribution are:

$$E(x) = \theta \tag{1.25}$$

$$Var(x) = \theta(1 - \theta) \tag{1.26}$$

In probability theory, a Bernoulli experiment is a series of repeated and identical random trials that are conducted independently, for which the outcome can take one of two values: success (the random variable takes the value 1) and failure (the random variable takes the value 0). For n such experiments, the process is also known as a Bernoulli experiment of n times.

The binomial distribution is a discrete probability distribution representing the number of successes in a Bernoulli experiment of n times. The parameter θ indicates the probability of success in each trial. A random variable x that follows the binomial distribution with parameters n and θ is typically denoted as $x \sim Bin(n, \theta)$. Such a distribution is shown in Fig. 1.4. Across n trials, the probability that the number of successes is exactly k is:

$$p(k|n, \theta) = \binom{n}{k} \theta^k (1 - \theta)^{n-k}, k = 0, 1, 2, \ldots, n \tag{1.27}$$

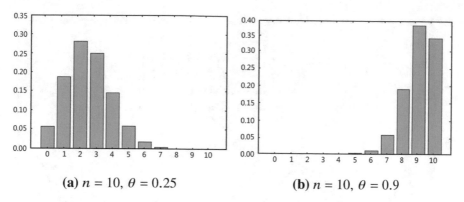

(a) $n = 10$, $\theta = 0.25$ **(b)** $n = 10$, $\theta = 0.9$

Fig. 1.4 Binomial distribution with the y-axis representing the probability

where $\binom{n}{k} = \frac{n!}{k!(n-k)!}$. If $x \sim Bin(n, \theta)$, then the expected value and variance of x are:

$$E(x) = n\theta \tag{1.28}$$

$$Var(x) = n\theta(1 - \theta) \tag{1.29}$$

The multinomial distribution is an extension of the binomial distribution. Consider throwing a K-side dice n times, where θ_j represents the probability that the dice lands with the jth side up and satisfies $S_K = \{\theta : 0 \leq \theta_j \leq 1, \sum_K \theta_j = 1\}$. Let $\mathbf{x} = (x_1, \ldots, x_K)$ represents the throwing result, where x_j indicates the number of the jth side up. Then, the probability density function of $\mathbf{x}$ is as follows:

$$p(\mathbf{x}|n, \boldsymbol{\theta}) = \frac{n!}{x_1! \cdots x_K!} \prod_{j=1}^{K} \theta_j^{x_j} \tag{1.30}$$

where θ_j is the probability of the jth side up.

The expected value and variance of multinomial distribution are:

$$E(x_j) = n\theta_j \tag{1.31}$$

$$Var(x_j) = n\theta_j(1 - \theta_j) \tag{1.32}$$

The beta distribution, typically denoted as $Be(\alpha, \beta)$, refers to a set of continuous probability distributions defined in the interval $(0, 1)$. The distribution has two parameters α and β, both of which are greater than zero. As shown in Fig. 1.5, the corresponding probability density function is:

$$p(x|\alpha, \beta) = \frac{\Gamma(\alpha + \beta)}{\Gamma(\alpha)\Gamma(\beta)} x^{\alpha-1}(1 - x)^{\beta-1}, 0 < x < 1 \tag{1.33}$$

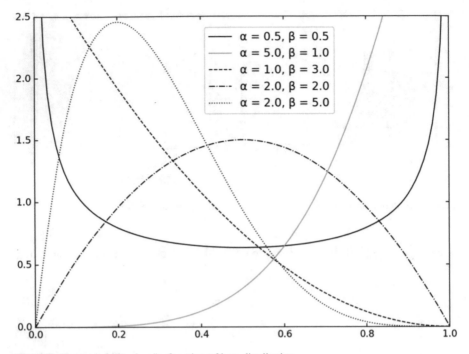

Fig. 1.5 The probability density function of beta distribution

The expected value and variance of the beta distribution are:

$$E(x) = \frac{\alpha}{\alpha + \beta} \tag{1.34}$$

$$Var(x) = \frac{\alpha\beta}{(\alpha + \beta)^2(\alpha + \beta + 1)} \tag{1.35}$$

The Dirichlet distribution refers to a group of continuous multivariable probability distributions. For a K-dimensional Dirichlet distribution, whose parameters satisfy $S_K = \{\alpha : 0 \leq \alpha_k \leq 1, \sum_K \alpha_k = 1\}$, the corresponding probability density function is:

$$p(\boldsymbol{x}|\boldsymbol{\alpha}) = \frac{1}{B(\boldsymbol{\alpha})} \prod_{i=1}^{K} x_i^{\alpha_i - 1} \tag{1.36}$$

where the function $B(\alpha)$ is a normalization factor, defined as:

$$B(\boldsymbol{\alpha}) = \frac{\prod_{i=1}^{K} \Gamma(\alpha_i)}{\Gamma(\sum_{i=1}^{K} \alpha_i)} \tag{1.37}$$

The expected value and variance of the Dirichlet distribution are:

$$E(x_i) = \frac{\alpha_i}{\sum_{i=1}^{K} \alpha_i} \tag{1.38}$$

$$Var(x_i) = \frac{\alpha_i(\sum_{i=1}^{K} \alpha_i - \alpha_i)}{(\sum_{i=1}^{K} \alpha_i + 1)(\sum_{i=1}^{K} \alpha_i)^2} \tag{1.39}$$

The Poisson distribution is primarily used to describe the probability distribution of a number of random events within a certain time period. The Poisson distribution has a corresponding probability density function:

$$p(x|\lambda) = \frac{\lambda^x e^{-\lambda}}{x!} \tag{1.40}$$

where λ is the average incidence of random events in unit time. Many real-world phenomena can be described by the Poisson distribution, such as the number of calls received by customer service personnel within a given interval and the number of natural disasters during a certain period of time. The expected value and variance of the Poisson distribution are both given by λ, as follows:

$$E(x_i) = \lambda \tag{1.41}$$

$$Var(x_i) = \lambda \tag{1.42}$$

The Gaussian distribution is a common continuous probability distribution, also known as the normal distribution. The probability density function of the Gaussian distribution is:

$$p(x|\mu, \sigma^2) = \frac{1}{\sqrt{2\pi}\sigma} \exp\{-\frac{1}{2\sigma^2}(x-\mu)^2\} \tag{1.43}$$

where μ and σ^2 are the expected value and variance of the Gaussian distribution, respectively. The special case for which $\mu = 0$ and $\sigma = 1$ is known as the standard normal distribution.

1.2.5 Exponential Family

The exponential family is an important set of distributions in probability theory. Distributions belonging to the exponential family take the following form:

$$p(x|\eta) = h(x)\exp\{\eta^{\mathrm{T}}T(x) - A(\eta)\} \tag{1.44}$$

where $T(x)$ is the sufficient statistic, η is known as the natural parameter, and $A(\eta)$ is the logarithmic form of the normalization factor. The partition function logarithm can be expressed as:

$$A(\eta) = \log \int h(x)\exp\{\eta^T T(x)\} \tag{1.45}$$

Many mathematical distributions belong to the exponential family, including the exponential distribution, the gamma distribution, the beta distribution, the Dirichlet distribution, the Bernoulli distribution, and the Poisson distribution.

Consider a random variable x with a distribution parameter θ. If θ can be calculated using only the information provided by $T(x)$, then $T(x)$ is referred to as the sufficient statistic of θ. The sufficient statistic can be expressed as:

$$p(\theta|T(x), x) = p(\theta|T(x)) \tag{1.46}$$

The above definition shows that in the exponential family, $T(x)$ is the sufficient statistic of η. The sufficient statistic possesses several key mathematical properties that can simplify calculations. For example, when handling data such as time series instead of storing the entire set of data, the corresponding values of the sufficient statistic can be stored, thereby providing the required information more concisely. The exponential family has the following properties:

- The first-order partial derivative of the logarithmic normalization factor $A(\eta)$ is the expected value of the sufficient statistic, that is:

$$\frac{\partial A}{\partial \eta^T} = E[T(x)] \tag{1.47}$$

- The second-order partial derivative of the logarithmic normalization factor $A(\eta)$ is the variance of sufficient statistic, that is:

$$\frac{\partial^2 A}{\partial \eta \partial \eta^T} = \text{Var}[T(x)] \tag{1.48}$$

Using the Bernoulli distribution as an example, we show how a distribution can be rewritten into the standard form of the exponential family. The Bernoulli distribution can be rewritten as:

$$
\begin{aligned}
p(x|\theta) &= \theta^x (1 - \theta)^{1-x} \\
&= \exp\left\{\log\left(\frac{\theta}{1-\theta}\right)x + \log(1 - \theta)\right\} \\
&= (1 - \theta)\exp\left\{x\log\left(\frac{\theta}{1-\theta}\right)\right\}
\end{aligned}
\tag{1.49}
$$

When compared with the definition of the exponential family distributions:

$$\eta = \log \frac{\theta}{1 - \theta} \qquad (1.50)$$

The parameters θ and η have the following relationship:

$$\theta = \frac{1}{1 + e^{-\eta}} \qquad (1.51)$$

1.2.6 Bayes' Theorem

Many problems require that $p(x|y)$ is calculated with the value of $p(y|x)$ already being known. Such problems can be solved using the Bayes' theorem, which characterizes the relationship between $p(y|x)$ and $p(x|y)$. Bayes' theorem can be expressed as follows:

$$p(x|y) = \frac{p(y|x)p(x)}{p(y)} \qquad (1.52)$$

$p(x)$ is known as a prior probability of x, because it does not influence the value of a variable y. Correspondingly, $p(x|y)$ is known as the conditional or a posterior probability of x given y. The probability $p(y|x)$ is known as the likelihood, and $p(y)$ as the marginal or a prior probability y which is typically used as the normalization factor. Thus, Bayes' theorem can also be expressed in the following manner: a posterior probability = (the likelihood × a prior probability) ÷ the normalized constant.

Figure 1.6 shows an example to explain the application of Bayes' theorem. The scenario features two containers: a container A contains three black balls and seven white balls; a container B contains eight black balls and two white balls. Suppose a ball is drawn randomly from one of the two containers and we know that the ball is black, we need to calculate the probability that the black ball was drawn

Fig. 1.6 Using Bayes' theorem to calculate the probability that a black ball is drawn from the container A

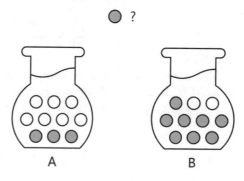

from the container A. Assuming that the drawing of the black ball is described by an event r, then the probability that a black ball is drawn is given by $p(r) = (3 + 8)/(3 + 7 + 8 + 2) = 11/20$. Now let us describe the drawing of the ball from the container A as an event a and the drawing of the ball from the container B as an event b. Given that the ball is drawn randomly from the two containers, $p(a) = p(b) = 1/2$. Drawing the black ball from the container A specifically has a conditional probability $p(r|a) = 3/10$. Correspondingly, drawing the black ball from the container B specifically has a conditional probability $p(r|b) = 8/10$. According to the Bayes' theorem, the probability that the black ball is drawn from the container A is:

$$p(a|r) = \frac{p(r|a) \cdot p(a)}{p(r)} = \frac{(3/10) \cdot (1/2)}{11/20} = 3/11 \tag{1.53}$$

1.2.7   Conjugate Distribution

If a posterior probability $p(x|y)$ and a prior probability of $p(x)$ of a random variable x belong to the same distribution family, then $p(x|y)$ and $p(x)$ are known as the conjugate distribution and $p(x)$ is known as the conjugate prior of the likelihood function $p(x|y)$. Conjugate distribution is a key characteristic of the exponential family.

• The beta distribution and binomial distribution are conjugate.
  Consider a random variable x with a prior:

$$p(x) = \frac{\Gamma(\alpha + \beta)}{\Gamma(\alpha)\Gamma(\beta)} x^{\alpha-1}(1 - x)^{\beta-1} \tag{1.54}$$

The likelihood function is:

$$p(y|x) = \binom{n}{k} p^k(1 - p)^{n-k} \tag{1.55}$$

According to the Bayes' theorem, the posterior is:

$$p(x|y) = \frac{\Gamma(\alpha + \beta + n)}{\Gamma(\alpha + k)\Gamma(\beta + n - k)} x^{\alpha+k-1}(1 - x)^{\beta+n-k-1} \tag{1.56}$$

This result shows that the posterior $p(x|y)$ belong to the beta distribution.
• The Dirichlet and multinomial distributions are conjugate.
  Consider a random variable x with a prior:

$$p(x; \alpha_1, \cdots, \alpha_K) = \frac{1}{B(\alpha)} \prod_{i=1}^{K} x_i^{\alpha_i - 1} \tag{1.57}$$

The likelihood function is:

$$p(y|x) = \frac{n!}{x_1! \cdots x_k!} p_1^{x_1} \times \cdots \times p_k^{x_k} \tag{1.58}$$

According to the Bayes' theorem, the posterior is:

$$p(x|y) = \frac{1}{B(\alpha + n_1 + \cdots + n_k)} \prod_{i=1}^{K} x_i^{\alpha_i + n_i - 1} \tag{1.59}$$

This result shows that the posterior $p(x|y)$ belong to the Dirichlet distribution.

These two examples show that for a given likelihood function, the difficulty of finding a posterior probability is dependent upon the selection of a prior distribution. Appropriate selection of the conjugate prior distribution and the likelihood function allows a posterior probability distribution to take the same form as a prior probability distribution, so that a closed-form solution of the former can be directly obtained.

1.2.8 Divergence

1.2.8.1 Kullback-Leibler Divergence

Kullback-Leibler (KL) divergence is a common measure that defines the difference between two probability distributions $p(x)$ and $q(x)$:

$$D_{KL}(p||q) = \sum_{x} p(x) \log \frac{p(x)}{q(x)} \tag{1.60}$$

The above definition shows that the KL divergence cannot be negative. Moreover, if and only if $q(x)$ and $p(x)$ are equal, $D_{KL}(p||q) = 0$. Note that KL divergence is not symmetric when measuring the difference between two probability distributions, that is, $D_{KL}(p||q) \neq D_{KL}(q||p)$.

1.2.8.2 Jensen-Shannon Divergence

Jensen-Shannon (JS) divergence provides a symmetrical solution to the asymmetry of KL divergence. JS divergence can take values between 0 and 1, and is defined as follows:

$$D_{JS}(p||q) = \frac{1}{2} D_{KL}\left(p||\frac{p+q}{2}\right) + \frac{1}{2} D_{KL}\left(q||\frac{p+q}{2}\right) \tag{1.61}$$

1.3   Bayesian Networks

Bayesian networks are tools used for uncertain reasoning. They play a key role in medical diagnosis, information retrieval, and automatic driving. Bayesian networks have graph structures that model the dependencies between variables. In combination with probability theory, Bayesian networks can be used to reason and predict outcomes. Topic models introduced in this book can be regarded as a subset of Bayesian networks. To facilitate learning later chapters, we introduce several important concepts and techniques related to Bayesian networks.

1.3.1   Representation

Mathematically, a Bayesian network is expressed as a tuple $(\mathcal{G}, \Theta)$, where $\mathcal{G}$ represents the network structure composed of variables based on the graph theory, and Θ represents the parameter information based on the probability theory. Figure 1.7a illustrates the basic elements based upon which $\mathcal{G}$ is constructed. Random variables are represented by circular nodes. Observed variables and latent variables are distinguished by shaded circular nodes and hollow circular nodes, respectively. Observed variables are those whose specific values can be obtained directly from the data set; latent variables are those whose specific values cannot be obtained directly from the data set. Directed edges indicate dependent relationships between random variables. Similar to the definition in graph theory, the variable at which the directed edge starts is referred to as the parent node and the variable at which the directed edge ends is referred to as the child node. The graph $\mathcal{G}$ has a directed acyclic structure composed of the basic elements shown in Fig. 1.7. The parameter information Θ contains the conditional probability distribution of each variable in $\mathcal{G}$. The joint probability distribution of all variables in a Bayesian network can be deduced from the graph structure $\mathcal{G}$ and the parameter information Θ.

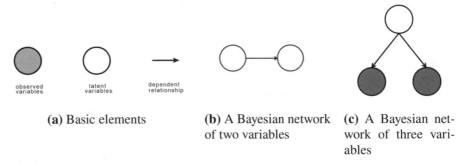

(a) Basic elements   (b) A Bayesian network of two variables   (c) A Bayesian network of three variables

Fig. 1.7 Examples of Bayesian networks

As an example, consider a Bayesian network composed of the variables $x_1, x_2, \ldots, x_n$, it has the following joint distribution:

$$P(x_1, x_2 \ldots, x_n) = \prod_{i=1}^{n} P(x_i | \pi(x_i)), \qquad (1.62)$$

where $\pi(x_i)$ represents the parent node set of x_i. When $\pi(x_i)$ is empty, $P(x_i | \pi(x_i))$ reduces to the marginal distribution $P(x_i)$. This equation shows that the joint distribution of all variables can be decomposed into the product of several conditional probabilities. These conditional probabilities can be flexibly modeled by appropriate probability distributions. For example, consider a discrete variable with a discrete parent node, a multinomial distribution can be used to represent the conditional probability. If the variable is instead continuous, the conditional probability could be modeled by a Gaussian distribution. To summarize, Bayesian networks has the following key characteristics:

- Each node is associated with a probability distribution.
- The directed edge represents dependency between variables.
- The joint distribution of all variables can be decomposed into the product of the conditional probabilities of each variable and its parents.

For a Bayesian network containing many variables, the graph structure G can be cumbersome. Substructures that repeatedly occur within G can be represented concisely using plate notation. Repeated substructures are drawn just once and enclosed within a rectangle and the number of repetitions is indicated by a number in the lower right corner of the rectangle. As shown in Fig. 1.8a, when plate notation is not used, the substructure of observed variable repeats many times. In contrast, the substructure is represented concisely by plate notation in Fig. 1.8b.

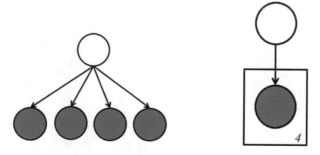

(a) Bayesian network without plate notation (b) Bayesian network using plate notation

Fig. 1.8 Plate notation

The graphical representation of a Bayesian network intuitively reveals the dependent relationships between variables but does not display any information contained within the parameter Θ. To illustrate the information contained within both $\mathcal{G}$ and Θ, a generative process is often used. Generative process describes the mathematical origin of the variables within the network in a storytelling manner.

Algorithm 1: The generative process of Näive Bayes classifier

---
1 for *each document d* **do**
2   draw a category label c from a probability distribution $p(c)$
3   **for** *each feature of the document d* **do**
4 | generate a feature f_i using the probability distribution $p(f_i|c)$
5   **end**
6 end

We use the Näive Bayes classifier as an example to illustrate the generative process. The Näive Bayes classifier is a common method of document classification. The classifier's graphical model is shown in Fig. 1.9. The observed variables are certain document features f_i, which are typically the words in the document. The latent variable c is the category. Directed edges are present only between the category node and nodes representing features. Algorithm 1 describes the generative process of the Näive Bayes classifier. Firstly, a category c is drawn for each document based on the probability $p(c)$. Following this, each feature f_i is generated according to the probability $p(f_i|c)$. As will become apparent in the following chapters, the Näive Bayes classifier can be regarded as an extremely simplistic topic model: each category is a topic and only one topic exists in a document. However, in reality, a document typically contains multiple topics and later chapters will introduce topic models in which it is assumed that multiple topics exist in a document.

Fig. 1.9 A graphical representation of the Näive Bayes classifier

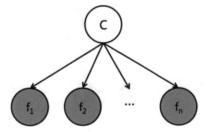

1.3.2 Conditional Independence

The primary purpose of the graph structure $\mathcal{G}$ is to explicitly model the relationships between variables. In this section, we describe how the graph structure can be analyzed to determine the conditional independence of the variables. This section provides readers with the means to quickly analyze the relationships between variables in Bayesian networks as well as that of constructing a Bayesian network according to given assumptions of independence. We start with simple structures consisting of two variables x and y connected by a third variable z. There are three possible structures: chain, fork, and collider.

Figure 1.10 shows the chain structure. We can obtain the joint probability distribution of x, y, and z: $p(x, y, z) = p(x)p(z|x)p(y|z)$. If all variables are unobserved, by integrating z we obtain $p(x, y) = p(x) \sum_z p(z|x)p(y|z) = p(x)p(y|x)$. If z is unknown, x and y are not independent, $p(x, y)$ cannot be decomposed into $p(x)p(y)$. If z is observed, the probability of x and y is $p(x, y|z) = \frac{p(x,y,z)}{p(z)} = \frac{p(x)p(z|x)p(y|z)}{p(z)} = p(x|z)p(y|z)$. Therefore, x and y are independent if z is observed.

Figure 1.11 shows the fork structure. According to the dependent relationship, we can get the joint probability distribution $p(x, y, z) = p(x|z)p(y|z)p(z)$. Assuming that all variables are unobserved, by integrating out z, we get $p(x, y) = \sum_z p(x|z)p(y|z)p(z)$. This formula cannot be decomposed into $p(x)p(y)$, so x and y are not independent under the condition that z is unobserved. Assuming that the variable z is observable, when z is known, the probability of x and y is $p(x, y|z) = \frac{p(x,y,z)}{p(z)} = p(x|z)p(y|z)$. Therefore, x and y are independent under the condition that z is observed.

Figure 1.12 shows the collider structure. The joint probability distribution is $p(x, y, z) = p(x)p(y)p(z|x, y)$. If all variables are unobserved, by integrating out z, we obtain $p(x, y) = p(x)p(y)$. If z is observable, the probability of x and y is $p(x, y|z) = \frac{p(x,y,z)}{p(z)} = \frac{p(x)p(y)p(z|x,y)}{p(z)}$. Hence, x and y are independent if z is unobserved.

The above discussion demonstrates the conditional independence between variables in simple Bayesian networks. For complex Bayesian networks, determining

Fig. 1.10 Chain

Fig. 1.11 Fork

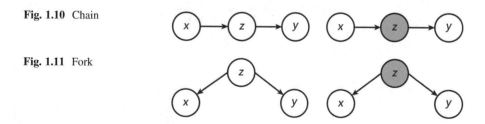

Fig. 1.12 Collider

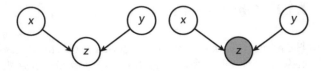

such independence relationship relies on the concept of d-separation, which provides a general approach to determine the independence between variables. Assume X, Y, and Z are three sets of nodes,[1] we want to determine whether X and Y are conditionally independent with respect to Z. To do this, we investigate all undirected paths connecting X and Y in $\mathcal{G}$. A path is blocked if one of the following conditions is met:

- There exists a chain or fork node $z \in Z$;
- There exists a collider node w, and neither w nor its descendants are in Z.

If all paths between X and Y are blocked, then X and Y are conditionally independent with respect to Z; otherwise, X and Y are not conditionally independent with respect to Z.

1.3.3 Uncertain Reasoning

The previous section has introduced methods to qualitatively analyze the relationships between variables from the graph structure $\mathcal{G}$ of a Bayesian network. We now consider how $\mathcal{G}$ and Θ can be used together to quantitatively analyze such relationships. Such quantitative analysis is often known as uncertain reasoning. More specifically, uncertain reasoning is defined as the process of calculating posteriori probabilities of query variables based on evidence variables.

In practice, evidence variables are typically observed variables, and query variables are typically latent variables. If x represents all observed variables and z represents all latent variables, the reasoning process calculates a posteriori of z:

$$p(z|x) = \frac{p(z, x)}{p(x)} = \frac{p(x, z)}{\int p(x, z)dz} \tag{1.63}$$

---

[1] A single node is a special case in which the number of nodes in the set is one.

Fig. 1.13 A chain Bayesian network of four latent variables

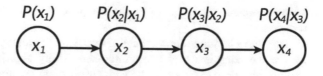

If the denominator has a closed-form solution, and we can conduct exact inference using the above equation. For example, the reasoning process for the Näive Bayes classifier can be expressed as follows:

$$p(C = c|f_1, \ldots, f_n) = \frac{p(c)p(f_1, \ldots, f_n|c)}{p(f_1, \ldots, f_n)}$$

$$= \frac{p(c) \prod_{i=1}^{n} p(f_i|c)}{\sum_{c'} p(f_1, \ldots, f_n, c')} \quad (1.64)$$

$$= \frac{p(c) \prod_{i=1}^{n} p(f_i|c)}{\sum_{c'} p(c') \prod_{i=1}^{n} p(f_i|c')}$$

For the Näive Bayes classifier, the latent variable C typically has a low dimension, and each observed variable is conditionally independent, and calculating the denominator $\sum_{c'} p(c') \prod_{i=1}^{n} p(f_i|c')$ is relatively simple. We can calculate the posterior probability of each category accurately.

For a Bayesian network with many latent variables, the complexity of integrating $\int p(x, z)dz$ increases exponentially with the number of variables. To reduce this complexity, we can exploit the independence between variables using more sophisticated algorithms. As an example, consider a chain Bayesian network shown in Fig. 1.13. The network has four latent variables, and $p(x_4)$ can be calculated as:

$$p(x_4) = \sum_{x_1, x_2, x_3} p(x_1, x_2, x_3, x_4) = \sum_{x_1, x_2, x_3} P(x_1)p(x_2|x_1)p(x_3|x_2)p(x_4|x_3)$$
$$(1.65)$$

If each variable is binary, then $4 + 8 + 16 = 28$ multiplication operations and $2 + 4 + 8 = 14$ addition operations are required to obtain $p(x_4)$. As shown in Eq. (1.65), some variables are only related to certain probabilities. For example, a variable x_1 is only related to $p(x_1)$ and $p(x_2|x_1)$). As such, the above equation can be further decomposed into:

$$p(x_4) = \sum_{x_3} p(x_4|x_3) \sum_{x_2} p(x_3|x_2) \sum_{x_1} p(x_1)p(x_2|x_1) \quad (1.66)$$

Given that x_2 is an observed variable, $p(x_4) = \sum_{x_3} p(x_4|x_3) \sum_{x_2} p(x_3|x_2)p(x_2)$. Therefore, only $4 + 4 = 8$ multiplication operations and $2 + 2 = 4$ addition operations are required to obtain $p(x_4)$ if x_2 and x_3 are eliminated by marginalization.

When compared with Eq. (1.65), the complexity of this calculation is substantially reduced.

The above example shows that decomposing the joint probability can effectively reduce the complexity of exact inference. Many algorithms for exact inference exploit this approach, such as the variable elimination algorithm and the clique tree algorithm. By localizing the calculation, these algorithms calculate the probability associated with a given node without the involvement of all variables, saving substantial calculation cost. However, both the variable elimination algorithm and the clique tree algorithm are only suitable for small-scale Bayesian networks with sparse graph structures. For large-scale Bayesian networks with dense connections, $\int p(x, z)dz$ typically has no closed-form solution. In this situation, we resort to approximate inference to estimate $p(x|z)$. Approximate inference improves efficiency at the expense of reducing accuracy. Common approximate inference algorithms include the Monte Carlo method and variational inference. As detailed in the following chapters, these two algorithms have their own advantages and disadvantages when applied to topic models.

1.3.4 Parameter Learning

Parameter learning refers to learning Θ by training data with a given graph structure $\mathcal{G}$. Such methods can be divided into two categories: point estimation and Bayesian estimation.

Point estimation methods calculate a single value, which is the best estimate of the distribution parameters that generate the trained data. Point estimation methods can be further divided into maximum likelihood estimation (MLE) and maximum a posteriori estimation (MAP). Assume that a data set is denoted as $\mathcal{D}$, the objective of MLE is to find a fixed value $\hat{\Theta}_{MLE}$ that satisfies:

$$\hat{\Theta}_{MLE} = \arg \max_{\Theta} p(\mathcal{D}|\Theta) \tag{1.67}$$

MAP additionally introduces a prior of Θ. The objective of MAP is to find a fixed value $\hat{\Theta}_{MAP}$ that satisfies:

$$\hat{\Theta}_{MAP} = \arg \max_{\Theta} p(\Theta|\mathcal{D}) = \arg \max_{\Theta} \frac{p(\mathcal{D}|\Theta)p(\Theta)}{p(\mathcal{D})} = \arg \max_{\Theta} p(\mathcal{D}|\Theta)p(\Theta) \tag{1.68}$$

The denominator $p(\mathcal{D})$ is a fixed value for Θ and can be ignored. Equations (1.67) and (1.68) further reveal the close relationship between MLE and MAP. MAP introduces a prior of parameters based on MLE; hence, MAP can be regarded as a regularized version of MLE. The presence of latent variables z impedes the use of MLE or MAP. The calculation of $p(\mathcal{D}|\Theta)$ (i.e., $\int P(\mathcal{D}|z, \Theta) f(z|\Theta)dz$) requires the integration of z and it is usually a difficult problem. In such a scenario, MLE and MAP can be solved using the expectation maximization (EM) algorithm, which will be introduced in Chap. 4.

Bayesian estimation is to find a distribution rather than a fixed value. Bayesian estimation regards Θ as a random variable and estimates its posterior:

$$p(\Theta|\mathcal{D}) = \frac{p(\mathcal{D}|\Theta) \cdot p(\Theta)}{p(\mathcal{D})} \tag{1.69}$$

Bayesian estimation is typically more complicated than MLE and MAP, because the denominator $p(\mathcal{D})$ cannot be omitted. The calculation of $p(\mathcal{D})$ involves the complex integral operation $p(\mathcal{D}) = \int_{\Theta} p(\mathcal{D}|\Theta)p(\Theta)d\Theta$. The difficulty of calculating $p(\mathcal{D})$ highlights the importance of conjugate prior. If a prior distribution and the distribution of the likelihood function are conjugate, the above integral can be calculated easily.

1.3.5 Structure Learning

The previous section describes the learning of the parameter Θ for a Bayesian network with a predefined structure $\mathcal{G}$. In this section, we investigate the learning of the network structure itself and this process is defined as structure learning, which includes two important concepts: model selection and model optimization.

Model selection uses a metric to evaluate the quality of a Bayesian network structure $\mathcal{G}$. A common metric is Bayesian information criterion (BIC), which is calculated as follows:

$$\text{BIC}(\mathcal{G}, \mathcal{D}) = \log P(\mathcal{D}|\hat{\Theta}, \mathcal{G}) - \frac{|\hat{\Theta}|}{2} \log N \tag{1.70}$$

where $\hat{\Theta}$ is the MLE estimate of the parameter Θ, $|\hat{\Theta}|$ is the number of parameters, and N is the number of samples in the trained data. Intuitively, BIC adds a penalty to the data likelihood in order to reduce the risk of overfitting. BIC does not depend on any prior information of parameters, so it can be applied to scenarios that has no prior. The other common metrics include the minimum description length (MDL) score and the Akaike information criterion (AIC).

Model optimization searches for the network structure having the highest score with respect to a given evaluation metric. The simplest model optimization is the exhaustive method: calculating the score of every possible network structure and the structure with the highest one is selected. This approach is only appropriate for very small networks. More practical approaches include those that use heuristic search algorithms, such as hill climbing. After starting from an initial structure, the hill climbing method iteratively modifies the current structure by using a search operation to obtain a series of candidate models and selects the best candidate model for the next iteration. Most topic models do not use structure learning, because their structures are predefined and only parameter learning is required. However, some topic models used in industry rely on structure learning and these models will be introduced in Chap. 2.

For a more detailed introduction of the concepts covered in this chapter, we refer readers to the relevant literature like [1–4].

References

1. Greub WH (2012) Linear algebra, vol 23. Springer Science & Business Media, New York
2. Koski T, Noble J (2011) Bayesian networks: an introduction. Wiley, New York
3. Mendenhall W, Beaver RJ, Beaver BM (2012) Introduction to probability and statistics. Cengage Learning, Boston
4. Strang G (2020) Linear algebra for everyone. Wellesley-Cambridge Press, Wellesley, MA, USA

Chapter 2
Topic Models

2.1 Basic Concepts

Topic models have attracted substantial attention from academia and industry. Researchers have designed various types of topic models that are applied to a wide range of tasks. In this chapter, we select some representative topic models for introducing the mathematical principles behind topic models. By studying this chapter, readers will have a deep understanding of the foundation of topic models, and the ability to select appropriate existing models or design brand-new models for their own scenarios. To facilitate understanding of the content within this book, we provide definitions for several key concepts:

- A **document** is a text article such as a composition, a webpage, a contract, etc.
- A **corpus** is a collection of multiple documents.
- A **word** is the basic unit in a document.
- A **token** is a word at a certain position within the document. The same word at a different position is regarded as a different token. For brevity, in this book, we refer to both words and tokens as words: readers can judge their specific meaning according to their contexts.
- A **vocabulary** is the collection of all unique words in a corpus.
- A **topic** is a collection of words and in each topic each word is typically associated with a numerical value to indicate its importance.
- A **topic model** is a model that extracts topics from a corpus and uses these topics to represent a set of documents.

Figure 2.1 illustrates the above concepts. Based on a given corpus and vocabulary, a topic model can discover topics that are composed of words in the vocabulary and assign each word a weight. Figure 2.1 shows three examples of topics. Topic 1 includes words such as "apple," "grape," and "orange," which have relative importance of 0.37, 0.13, and 0.12, respectively. Similarly, topic 2 includes the words "apple," "iPad," and "iPhone," with importance of 0.45, 0.20, and 0.08,

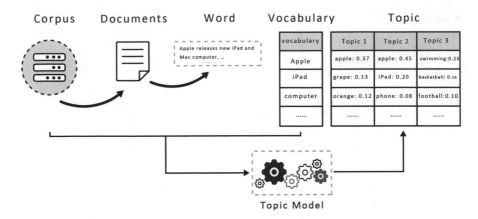

Fig. 2.1 The concepts of topic models

respectively. The topics can be used to solve two classic problems in text mining: synonymy and polysemy. Synonymy describes different words that have identical or similar meanings. For example, a topic model could find that "apple" and "grape" belong to the same topic, indicating that the two words have a semantic relationship. Polysemy refers to the phenomenon that a single word that has multiple different meanings. For example, a topic model could find that the word "apple" has multiple semantics because it belongs to a topic of electronic products as well as a topic of fruits. The topics transcend the low-level literal representations and make information processing possible at the semantic level.

2.2   Latent Semantic Analysis

Latent semantic indexing [4] is a method used to improve the quality of information retrieval by mining the latent semantic information from a text corpus. Given its widespread use in text information processing, this method is often referred to as latent semantic analysis (LSA).

The fundamental technique of LSA is singular value decomposition. Consider a vocabulary containing $|V|$ words and a corpus containing $|D|$ documents. We can construct a word-document matrix $X \in \mathbb{R}^{|V| \times |D|}$. The value of each element in X is either 0 or 1. 0 indicates that the corresponding word does not appear in the document; 1 indicates that the corresponding word appears in the document. We now apply the singular value decomposition to decompose X into three matrices:

$$X = U \Sigma V^T \tag{2.1}$$

where $U \in \mathbb{R}^{|V| \times m}$, $\Sigma \in \mathbb{R}^{m \times m}$, $V \in \mathbb{R}^{m \times |D|}$. U and V can be intuitively interpreted as word-topic and document-topic matrices, respectively. U reveals the importance of the words in each topic while V^T produces the representations of documents in the topic space. Note that Σ is a diagonal matrix and values along its diagonal are referred to as singular values. LSA truncate small singular values in Σ and its corresponding rows and columns in U and V, in order to produce a representation of U and V with reduced dimensions. Consider retaining only the first K ($K < m$) singular values, we truncate the original matrix Σ to produce the matrix $\Sigma_K \in \mathbb{R}^{K \times K}$, containing only the first K singular values, and truncate U to $U_K \in \mathbb{R}^{|V| \times K}$ and $V_K^T \in \mathbb{R}^{K \times |D|}$. In LSA, K is regarded as a hyperparameter and can be set manually.

Figure 2.2 provides a simple example LSA. Figure 2.2a shows a word-document matrix X. By applying LSA to X, we obtain a word-topic matrix U (Fig. 2.2b) and a document-topic matrix V^T (Fig. 2.2c). The word-topic matrix shows the corresponding importance of each word within topics. For example, in the first topic $t1$, "singer" and "instrument" have importance of 0.41 and 0.38, respectively. Similarly, the document-topic matrix shows the corresponding importance of each topic within documents. For example, in the first document, topic $t1$ has an importance of 0.61 and topic $t2$ has an importance of -0.38.

LSA is widely used in the field of information retrieval. The obtained matrices U_K, Σ_K and V_K^T are used to reconstruct a new word-document matrix $X_K = U_K \Sigma_K V_K^T$. X_K provides semantic connections between the document and words that are not present in the document, improving the performance of document classification and document clustering.

However, LSA retains several inherent problems. (1) The optimal setting of the hyperparameter K remains an open problem. (2) Singular value decomposition can result in excessive computation requirements and high memory consumption. (3) The word-topic and document-topic matrices have both positive and negative elements, and the importance of a corresponding word or topic cannot be determined intuitively from these numerical values, making LSA results difficult to interpret.

Document Word	d1	d2	d3	d4	d5
singer	1	0	1	0	0
instrument	0	1	1	1	0
music	1	1	0	0	1
guitar	1	0	1	0	0
piano	1	0	1	0	1
record	0	1	0	1	0

Topic Word	t1	t2	t3	t4
singer	0.41	-0.25	0.32	-0.39
instrument	0.38	0.58	0.39	0.30
music	0.44	0.07	-0.78	-0.26
guitar	0.41	-0.25	0.32	-0.39
piano	0.52	-0.32	0.11	0.69
record	0.18	0.64	-0.05	-0.21

Topic Document	t1	t2	t3	t4
d1	0.61	-0.38	0.18	-0.57
d2	0.34	0.66	0.31	-0.27
d3	0.59	-0.12	-0.64	0.32
d4	0.19	0.62	-0.23	0.14
d5	0.33	-0.12	0.63	0.68

(a) Word-document matrix   **(b)** Word-topic matrix   **(c)** Document-topic matrix

Fig. 2.2 An example of LSA

2.3   Probabilistic LSA

To alleviate the above problems of LSA, probabilistic LSA (PLSA) [5] is proposed by researchers. PLSA assumes that each document in the corpus has a multinomial distribution of the latent topics. For example, consider a bioinformatics document that contains two topics: "computer" and "biology." The former topic accounts for 60% and the latter accounts for 40%. PLSA also assumes that each topic is a multinomial distribution of words. The topic about computer is composed of the words: computer, speed, memory, genetics and protein, with proportions of 80%, 10%, 5%, 5% and 0%. The topic about biology is composed of the same set of words with proportions of 0%, 0%, 0%, 80%, and 20%.

We use d, z, and w to denote documents, topics and words. A document is a set of words $(w_1, w_2, \ldots, w_n)$, where w_n is the nth word in the document. A corpus $\mathcal{D}$ is a set of documents $(d_1, d_2, \ldots, d_m)$, where d_m is the mth document in the corpus. Since PLSA belongs to Bayesian network, we can use Bayesian network techniques to explain the mathematical principles of PLSA. Figure 2.3 illustrates plate notation of PLSA. Given that documents and words can be observed, d and w are observed variables. The topic cannot be observed and z are latent variables. The variable M, shown in the lower right corner of the outer rectangle, represents the number of documents in the corpus. The variable N, shown in the lower right corner of inner rectangle, indicates that the document contains N words. Therefore, there are M variables d in PLSA. Each variable d generates N latent variables z, and each variable z generates a word w. Some work annotates the parameter information through the plate notation. With the parameters $p(d)$, $p(z|d)$, and $p(w|z)$, the plate notation provides a more explicit demonstration of the mathematical relationships between variables.

Algorithm 1 describes the generative process of PLSA. The variable $p(d)$ represents the probability that a document d occurs in the data set and it is usually treated as a fixed value in practical applications. The variables $p(z|d)$ and $p(w|z)$ represent the probability that a topic z occurs in the document d and the probability that a word w occurs in topic z. When using PLSA, the topic distribution within the document and the word distribution within the topic are assumed to be multinomial. As such, $p(z|d)$ and $p(w|z)$ are modeled as multinomial distributions. According

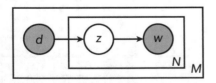

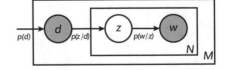

(a) Graphical model of PLSA without parameter information

(b) Graphical model of PLSA with parameter information

Fig. 2.3 Graphical models of PLSA without/with parameter information

---

Algorithm 1: The generative process of PLSA

---
1 for *each document d in D* **do**
2   generate the document d with a probability $p(d)$;
3   **for** *each position in d* **do**
4   generate a topic z with a probability $p(z|d)$;
5   generate a word w with a probability $p(w|z)$;
6   **end**
7 end

---

to these assumptions, the log likelihood of all documents in a corpus is given by:

$$\mathcal{L} = \sum_{d \in \mathcal{D}} \sum_{w \in \mathcal{V}} n(d, w) \log p(d, w) \tag{2.2}$$

where $n(d, w)$ represents the number of a word w in the document d, and $p(d, w)$ represents the joint probability of d and w. Given that $p(d, w) = p(d) \sum_{z \in \mathcal{Z}} p(w|z) p(z|d)$, the above equation can be further shown as follows:

$$\mathcal{L} = \sum_{d \in \mathcal{D}} \left[n(d) \log p(d) + \sum_{w \in \mathcal{V}} n(d, w) \log \sum_{z \in \mathcal{Z}} p(w|z) p(z|d) \right] \tag{2.3}$$

where $n(d) = \sum_{w \in \mathcal{V}} n(d, w)$. Since $\mathcal{L}$ contains the latent variable z, the expectation maximization algorithm is typically used to obtain the point estimates of $p(z|d)$ and $p(w|z)$. This algorithm will be introduced in detail in Chap. 4.

Similar to LSA, PLSA requires that the number of topics is predefined. However, PLSA has more solid probabilistic foundation. Therefore, the parameters of PLSA are intuitively interpretable. Figure 2.4 shows topics generated by PLSA, with the words in each topic z sorted based on $p(w|z)$ in descending order. The word weights of PLSA topics are all positive. Hence, the weights clearly indicate the importance of each word within the topic, providing PLSA with a greater degree of interpretability than LSA.

PLSA has some inherent disadvantages. It regards the topic distribution $p(z|d)$ of each document as parameter, without further assumption on how the distribution should be generated. As such, PLSA is incomplete as a generative model, and its parameters increase linearly with the number of documents in the training corpus. This over-reliance on the training corpus gives PLSA a problem with overfitting. Moreover, given that the learned parameter $p(z|d)$ represents the topic distribution of a document in the training data, PLSA cannot naturally obtain the topic distribution for a document beyond the training data set.

Fig. 2.4 Examples of PLSA topics

Topic 1	Topic 2	Topic 3
hip hop: 0.457	Apple: 0.374	clothes: 0.500
music style: 0.223	Samsung: 0.239	color: 0.202
psychedelia: 0.101	power: 0.155	good looking: 0.054
electronic: 0.051	Huawei: 0.113	design: 0.031
music: 0.032	screen: 0.022	suited: 0.013
rock:0.011	data cable: 0.021	style: 0.011
composition: 0.006	new arrival: 0.010	weaving: 0.010
pop:0.003	iPad: 0.007	sports style: 0.005
song: 0.001	processor: 0.002	fashion: 0.003
hit song: 0.001	release: 0.002	detail: 0.001

2.4   Latent Dirichlet Allocation

To overcome the disadvantages of PLSA, researchers proposed the latent Dirichlet allocation (LDA) [3]. Whereas PLSA regards the topic distribution $p(z|d)$ as a parameter, LDA regards $p(z|d)$ as a random variable θ and assumes that θ is generated by a Dirichlet prior. In this manner, the number of parameters does not increase with the size of the corpus increases. Given that a topic distribution θ is generated from a Dirichlet distribution, θ can be calculated naturally for documents beyond the training data.

There are two variants of LDA in the existing literature: the smoothed variant and the non-smoothed variant. Figure 2.5 shows graphical models of both variants constructed using plate notation. The two variants share some common features. The outermost rectangle represents the document. The number M, shown in the lower right corner of the rectangle, indicates a total of M repetitions, since there are M documents in the corpus. The inner rectangle contains variables associated with each word. The number N, shown in the lower right corner, indicates a total of N repetitions, which means that there are N words within the document. In both variants, each document is assumed to have a topic distribution, denoted by θ, and each topic takes the form of a multinomial distribution across all words in the vocabulary, denoted by ϕ.

The primary difference between the two LDA variants is the topic-word distribution ϕ. The non-smoothed variant of LDA regards ϕ as an $\mathbb{R}^{K \times V}$ matrix, with

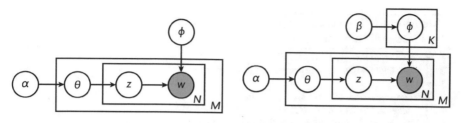

(a) Non-smoothed variant of LDA   **(b)** Smoothed variant of LDA

Fig. 2.5 Graphical models of the non-smoothed and the smoothed variants of LDA

each $\phi_{ij} = p(w_j|z_i)$. This matrix is treated as a parameter rather than a variable. In contrast, the smoothed variant regards ϕ as a random variable generated by a Dirichlet prior. As vocabularies are typically large and some words may not appear in the training corpus, when using the non-smoothed variant of LDA, the generated topic will assign any word that is not in the training data a probability of zero. This problem is effectively solved by the smoothed variant of LDA, which assigns predefined probability to the words that do not exist in the training data. This book focuses on the smoothed LDA variant. Unless otherwise specified, LDA refers to the smoothed LDA variant. Similar to LSA and PLSA, LDA uses a predefined value K to determine the number of topics.

Algorithm 2: The generative process of LDA

---

1 for *each topic $k \in \{1, 2, \ldots, K\}$* **do**
2 $\quad$ generate a topic-word distribution $\phi_k \sim$ Dirichlet(β);
3 end
4 for *a document d_m, $m \in \{1, 2, \ldots, M\}$* **do**
5 $\quad$ generate a document-topic distribution $\theta_m \sim$ Dirichlet(α);
6 $\quad$ **for** *each position in d_m $n \in \{1, 2, \ldots, N_m\}$* **do**
7 $\quad\quad$ generate a topic $z_{m,n} \sim$ Multinomial(θ_m)
8 $\quad\quad$ generate a word $w_{m,n} \sim$ Multinomial($\phi_{z_{m,n}}$)
9 $\quad$ **end**
10 end

Based on the above assumptions, Algorithm 2 describes the generative process of LDA, where α and β are hyperparameters, θ_m is the topic distribution for a document d_m, and ϕ_k is the word distribution for a topic k, $\Phi = \{\phi_k\}_{k=1}^{K}$ and $\Theta = \{\theta_m\}_{m=1}^{M}$. As the algorithm shows, ϕ_k is a global variable, θ_m is a document-level variable, and w and z are word-level variables. The LDA generative process traverses every position of each document within the corpus. At each position, the document-topic distribution generates a topic, and then the topic-word distribution generate a specific word.

Algorithm 3: The generative process of LDA (The document length generated by Poisson distribution)

---

1 **for** *each topic $k \in \{1, 2, \ldots, K\}$* **do**
2 $\quad$ generate a topic-word distribution $\phi_k \sim$ Dirichlet(β);
3 **end**
4 **for** *a document d_m, $m \in \{1, 2, \ldots, M\}$* **do**
5 $\quad$ generate the length of a document $N_m \sim$ Poission(ξ)
6 $\quad$ generate a document-topic distribution $\theta_m \sim$ Dirichlet(α);
7 $\quad$ **for** *each position in d_m $n \in \{1, 2, \ldots, N_m\}$* **do**
8 $\quad\quad$ generate a topic $z_{m,n} \sim$ Multinomial(θ_m)
9 $\quad\quad$ generate a word $w_{m,n} \sim$ Multinomial($\phi_{z_{m,n}}$)
10 $\quad$ **end**
11 **end**

Note that some authors may describe the generative process of LDA in a different way, which is shown in Algorithm 3. Compared with Algorithm 2, Algorithm 3 includes the number of the positions N_m within each document. N_m is typically omitted since it is independent of other observed variables. In this book, we use Algorithm 2 as the standard generative process of LDA. According to the generative process of LDA, the joint probability distribution of observed variables and latent variables can be written as:

$$p(w, z, \theta_m, \Phi | \alpha, \beta) = \underbrace{p(\Phi | \beta)}_{\text{Topic Plate}} \; \overbrace{p(\theta_m | \alpha) \underbrace{\prod_{n=1}^{N_m} p(w_{m,n} | \phi_{z_{m,n}}) p(z_{m,n} | \theta_m)}_{\text{Word Plate}}}^{\text{Document Plate}} \quad (2.4)$$

where $\alpha = (\alpha_1, \ldots, \alpha_K)$ and $\beta = (\beta_1, \ldots, \beta_V)$. Each element in above equation corresponds with one of the variables shown in Fig. 2.5. The relevant variables are z, θ_m and Φ. The posteriori of these variables is as follows:

$$P(z, \theta_m, \Phi | w, \alpha, \beta) = \frac{P(w, z, \theta_m, \Phi | \alpha, \beta)}{P(w | \alpha, \beta)} \quad (2.5)$$

However, the calculation of the denominator $P(w | \alpha, \beta)$ requires that all latent variables are integrated. It is difficult due to the coupling between these latent variables. One possible solution is to integrate θ_m and Φ by sampling the posterior of z to estimate θ_m and Φ. This is known as the Markov Chain Monte Carlo method and will be covered in Chap. 5. Another approach is to decouple the latent variables and utilize variational distribution formed by each independent distribution to approximate the original conditional distribution of latent variables. This is known as the variational inference method and will be covered in Chap. 6.

Fig. 2.6 Examples of LDA topics

Topic 1	Topic 2	Topic 3
apple: 0.522	Apple: 0.457	swimming: 0.231
orange: 0.142	iPad: 0.200	basketball: 0.166
banana: 0.110	phone: 0.089	football: 0.103
size: 0.044	Samsung: 0.046	badminton: 0.101
fresh: 0.032	Cook: 0.022	taekwondo: 0.095
pear: 0.021	new release: 0.021	bowling: 0.072
peach: 0.012	computer: 0.009	fitness: 0.062
mango: 0.008	Huawei: 0.002	table tennis: 0.067
supermarket: 0.005	power: 0.001	baseball: 0.053
buy: 0.004	headphone: 0.001	tennis: 0.050

Figure 2.6 shows examples of LDA topics. Mathematically, LDA generates the topic distribution of a document using the Dirichlet prior. Compared with PLSA, for which the topic distribution of a document is set directly as a model parameter, LDA uses a more natural method to infer the topic distribution of a new document.

2.5 SentenceLDA

None of the above topic models (i.e., LSA, PLSA, and LDA) is capable of modeling word orders within documents, since these models are based on the bag-of-words assumption. Therefore, they can only model the co-occurrence of words at the document level. However, the co-occurrence of words within the same sentence tends to suggest stronger semantic relevance. An assumption that the words within the same sentence are generated from the same topic can further increase the interpretability of the topics.

SentenceLDA [1, 8] assumes that all words within a sentence are generated by the same topic. This assumption is highly consistent with human intuition. Figure 2.7 illustrates the graphical model of SentenceLDA. α and β are the hyperparameters. The outermost rectangle represents the document with the document variables contained within. The number M indicates that there are M documents in the corpus.

Fig. 2.7 Graphical model of
SentenceLDA

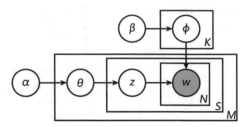

The number S indicates that there are S sentences in the document. The number N indicates that there are N words in the sentence. By comparing the graphical models of LDA and SentenceLDA, we can see that the difference is the presence of the inner rectangle. This rectangle contains the word variables and allows multiple words to be generated for the same topic z. In LDA, each variable z corresponds to a word w. However, in SentenceLDA, each variable z corresponds to multiple words.

Algorithm 4: The generative process of SentenceLDA

---

1 **for** *each topic* $k \in \{1, 2, \ldots, K\}$ **do**
2 $\quad$ generate a topic-word distribution $\phi_k \sim$ Dirichlet (β);
3 **end**
4 **for** *each document* d_m, $m \in \{1, 2, \ldots, M\}$ **do**
5 $\quad$ generate a document-topic distribution $\theta_{d_m} \sim$ Dirichlet (α);
6 $\quad$ **for** *each sentence in* d_m, $s \in \{1, 2, \ldots, S_m\}$ **do**
7 $\quad\quad$ generate a topic $z_{m,s} \sim$ Multinomial $(\theta_{d,m})$
8 $\quad\quad$ **for** *each position in* s, $n \in \{1, 2, \ldots, N_s\}$ **do**
9 $\quad\quad\quad$ generate a word $w_{m,s,n} \sim$ Multinomial$(\phi_{z_{m,s}})$
10 $\quad\quad$ **end**
11 $\quad$ **end**
12 **end**

Algorithm 4 describes the generative progress of SentenceLDA. Firstly, a document-topic distribution θ_{d_m} is generated for each document. For each sentence, a topic $z_{d_m s}$ is generated according to θ_{d_m} and a word is generated for each position in the sentence, according to the topic-word distribution $\phi_{z_{d_m s}}$. According to these assumptions, the joint probability distribution of observed and latent variables is:

$$p(\boldsymbol{w}, \boldsymbol{z}, \boldsymbol{\theta_m}, \Phi | \alpha, \beta) = \underbrace{p(\Phi|\beta)}_{\text{Topic Plate}} \, p(\boldsymbol{\theta_m}|\alpha) \overbrace{\underbrace{\prod_{s=1}^{S_m} \left[p(z_{m,s}|\boldsymbol{\theta_m}) \overbrace{\prod_{n=1}^{N_s} p(w_{m,s,n}|\phi_{z_{m,s}})}^{N_s} \right]}_{\text{Sentence Plate}}}^{\text{Document Plate}}$$

Word Plate

$$(2.6)$$

Fig. 2.8 Examples of
SentenceLDA topics

Topic 1	Topic 2	Topic 3
fruit: 0.058	phone :0.072	kitten: 0.062
apple: 0.044	Xiaomi: 0.059	puppy: 0.048
grape: 0.026	iPhone:0.056	hedgehog: 0.043
banana: 0.016	Apple: 0.040	duckling: 0.031
fresh: 0.015	Samsung: 0.031	critters: 0.029
strawberry: 0.015	review: 0.028	chick: 0.026
kiwi: 0.013	Huawei: 0.022	dog: 0.017
lychee: 0.013	price: 0.020	rabbit: 0.015
mango: 0.013	Red Mi: 0.016	monkey: 0.014
watermelon: 0.012	Honor: 0.016	mouse: 0.011

where S_m represents the number of sentences in a document m, and N_s represents the number of words in a sentence s.

Figure 2.8 shows examples of some SenteneceLDA topics. The topics have the same format as those of LDA. SentenceLDA models the sentence structure in documents and the topic contents are more coherent. Hence, SentenceLDA is often more effective than LDA in tasks such as text classification.

2.6 Topic over Time Model

The topic models described in the above sections assume that topics are static and do not change over time. However, the dynamic changes of topic intensity are very common and capturing these changes is meaningful for many downstream applications. Many text data such as blog posts, tweets, and search engine logs contain timestamp information, based on which researchers proposed the topic over time (TOT) model [14]. TOT detects changes of the topic intensity over time. Figure 2.9 shows the graphical model of TOT. The major difference between TOT and LDA is the observed timestamp variable t, which obeys the beta distribution defined by the parameter Ψ. This beta distribution quantifies the strength of a topic across time.

Algorithm 5: The generative process of TOT

---

1 for *each topic* $k \in \{1, 2, \ldots, K\}$ **do**
2 | generate a topic-word distribution $\phi_k \sim$ Dirichlet(β);
3 end
4 for *a document* d_m, $m \in \{1, 2, \ldots, D\}$ **do**
5 | generate a document-topic distribution $\theta_m \sim$ Dirichlet(α);
6 | **for** *each position in* d_m, $n \in \{1, 2, \ldots, N_m\}$ **do**
7 | | generate a topic $z_{m,n} \sim$ Multinomial(θ_m)
8 | | generate a word $w_{m,n} \sim$ Multinomial($\phi_{z_{m,n}}$)
9 | | generate a timestamp $t_{m,n} \sim$ Beta($\psi_{z_{m,n}}$);
10 | **end**
11 end

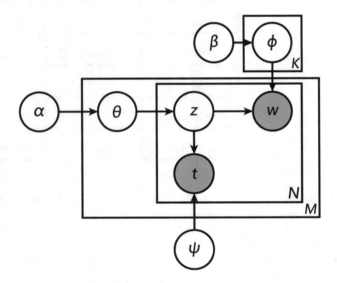

Fig. 2.9 Graphical model of TOT

Algorithm 5 presents the generative process of TOT. Firstly, a document-topic distribution θ_m is generated for each document d_m. For each position in the document, a topic z is generated from the document's topic distribution θ_{d_m}. A word is generated from the topic-word distribution ϕ_z. Next, the distribution ψ_z is used to generate a timestamp t for the word. The joint distribution of words, topics and timestamps in a document d_m can be calculated as follows:

$$p(\boldsymbol{w}, \boldsymbol{t}, z, \boldsymbol{\theta_m}, \Phi, \Psi | \alpha, \beta) = \underbrace{p(\Phi|\beta)}_{\text{Topic Plate}}$$

$$\overbrace{p(\boldsymbol{\theta_m}|\alpha) \underbrace{\prod_{n=1}^{N_m} p(w_{m,n}|\phi_{z_{m,n}})p(t_{m,n}|\psi_{z_{m,n}})p(z_{m,n}|\boldsymbol{\theta_m})}_{\text{Word Plate}}}^{\text{Document Plate}} \qquad (2.7)$$

Compared with LDA, the primary difference of TOT is the addition of $p(t_{m,n}|\Psi_{z_{m,n}})$, which provides the model with the ability to model topic strength over time. The two parameters of the beta distribution ψ_k can be estimated by timestamp samples of a topic k. Visualizing the shape of ψ_k presents the intensity changes of the topic k over time. Figure 2.10 shows examples of TOT topics used to describe the Mexican-American War, which took place during the late 1840s.

In addition to TOT, there are some topic models that are capable of handling time information and discovering the evolution of topics over time. For example, the dynamic topic model [2] divides the corpus into time slices and uses a Gaussian distribution to model the relationship between the topic-word distributions of adjacent time slices, thereby modeling the migration of the topics over time. The method used by TOT to handle timestamp information can also be used to process spatial information. Some topic models [6, 7, 13] replaces the beta distribution with a Gaussian distribution to model the longitude and latitude information related to a topic, thereby modeling the distribution of topics over the geographic locations.

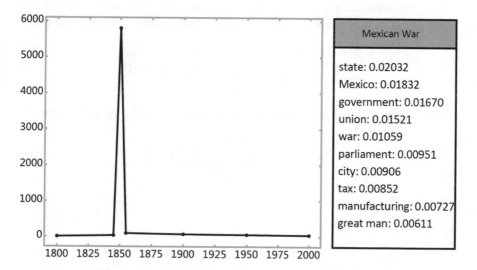

Fig. 2.10 An examples of TOT topic

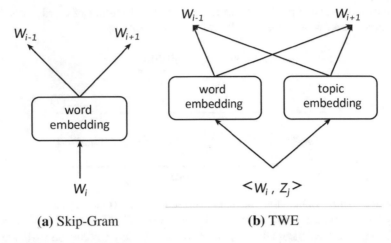

(a) Skip-Gram   **(b)** TWE

Fig. 2.11 Skip-Gram and TWE

2.7 Topical Word Embedding

Word embedding has become a popular technique in recent years and is widely used in many applications. Researchers have proposed methods to introduce word embedding into topic models. In this section, we discuss topical word embedding (TWE).[1] Training TWE has two major steps: (1) A topic z_i is sampled for each word w_i using the MCMC sampling algorithm.[2] (2) Based upon tuples $< w_i, z_i >$, the Word2Vec algorithm [11] is used to obtain the topic and word embedding.

For training word embedding, we select the Skip-Gram architecture . As shown in Fig. 2.11a, Skip-Gram learns the embedding representation of each word and uses that representation to predict the contextual words. For a document d containing words $\{w_1, \ldots, w_N\}$, the goal of Skip-Gram is to minimize the objective function:

$$\mathcal{L}(d) = \frac{1}{N} \sum_{i=1}^{N} \sum_{-n \leq c \leq n, c \neq 0} \log p(w_{i+c}|w_i) \tag{2.8}$$

where n represents the size of the context window predicted by the target word.

$$p(w_{i+c}|w_i) = \frac{\exp(e(\mathrm{w_c}) \cdot e(\mathrm{w_i}))}{\sum_{w_i \in W} \exp(e(\mathrm{w_c}) \cdot e(\mathrm{w_i}))} \tag{2.9}$$

---

[1] We choose the TWE-1 model [10].

[2] MCMC sampling algorithm will be explained in detail in Chap. 5. In this chapter, readers only need to know that the MCMC sampling algorithm annotates each word with a topic.

The above equation calculates the probability that the word w_{i+c} occurs within the context window of a word w_i; $e(w_i)$ represents the word embedding representation of w_i. As shown in Fig. 2.11b, TWE modifies Skip-Gram, and the topic of each target word is treated as a pseudo word and is also used to predict the words within context. Therefore, for TWE, the objective function is:

$$\mathcal{L}(d) = \frac{1}{N} \sum_{i=1}^{N} \sum_{-n \leq c \leq n, c \neq 0} \log P(w_{i+c}|w_i) + \log P(w_{i+c}|z_i) \qquad (2.10)$$

During the first stage of the training process, the embedding of each word is learned according to the Skip-Gram. Then, the word embedding is fixed and the topic embedding is optimized with regard to Eq. (2.10). TWE uses LDA for the first stage of training. However, given that the two stages are clearly decoupled, LDA can be replaced by alternative topic models during the first stage.

The major benefit of introducing word embedding into topic models is to obtain semantic information on low-frequency words. Traditional topic models assume that the topic is a multinomial distribution over the words. As such, the majority of the weight is occupied by high-frequency words and the role of low-frequency words is limited. The introduction of word embedding allows the embedding of words and topics to be obtained simultaneously and the relevance between topics and words to be evaluated from another perspective. This partially solves the problem that topics are dominated by high-frequency words. Figure 2.12 shows two examples. The topics based on multinomial distribution are overwhelmingly occupied by high-frequency words. For example, "music" and "opera" have frequencies of 2,057,265 and 90,002, respectively. When using word embedding, the words "ensemble" and "erhu" have frequencies of 3267 and 2626, respectively. By obtaining topics using both word embedding and multinomial distribution, semantic information of both high-frequency and low-frequency words can be successfully modeled.

	Multinomial Distribution	Embedding
Topic 1	music (2057265) piano (215204) concert (202814) play (117898) instrument (113147) art (2862128) performance (1162718) orchestra (96819) chorus (109177) opera (90002)	play (117898) violin (36597) percussion (10766) performer (4837) symphony (26829) solo (22460) chamber music (3295) concert (4556) ensemble (3267) erhu (2626)
Topic 2	collocation (946681) shirt (154773) coat (174422) black (955802) jacket (147491) white (724618) jeans (97279) fashion (1313866) skirt (112108) suit (118316)	dress (92139) shirt (154773) jeans (97279) fashion sense (2030) skirt (18696) smock (1806) sling (28945) high waist (9363) jumpsuit (5781) long-styled (14322)

Fig. 2.12 Topics based on multinomial distribution and based on embedding (The numbers indicate the frequency of the word)

2.8 Hierarchical Topic Models

The topic models introduced above provide only one layer of topical abstraction, and only the relationship between topics and words is modeled. Many real-life applications require multi-layer topical abstraction in order to model the hierarchical relations between topics. Consider a corpus that contains four topics, "cooking," "health," "insurance," and "medicine." Intuitively, "cooking" and "health" are semantically related, and "health," "insurance," and "medicine" are semantically related. This kind of semantic relations can be discovered with hierarchical topic models. In this section, we describe two hierarchical topic models: pachinko allocation model (PAM) and Rephil.

2.8.1 Pachinko Allocation Model

The pachinko allocation model (PAM) [9] discovers the hierarchical structures between topics. The PAM consists of a network with leaf nodes corresponding to words and the other layers of nodes corresponding to topics. The nodes of the upper layer can be regarded as a semantic abstraction of the lower layer nodes. PAM can discover topic structures with any number of layers. For brevity, we illustrate its mathematical foundation of PAM using a four-layer one. A four-layer PAM consists of a root node, two topic layers, and a layer of leaf nodes. The topic nodes directly connecting the root node is known as the super-topics, and the topic nodes directly connecting the leaf nodes is known as the sub-topics.

Algorithm 6 presents the generative process of a four-layer PAM. The generation of a document d_m begins with the root node. The topic of each lower layer is iteratively generated and a topic path is formed: $L_w = < z_{w1}, z_{w2}, z_{w3} >$. Then a word is sampled according to the corresponding topic-word distribution of the node z_{w3} in the path. This process is repeated until all the words in d_m are generated. As shown in Fig. 2.13, a four-layer PAM requires multiple additional variables to

Fig. 2.13 Graphical model of a four-layer PAM

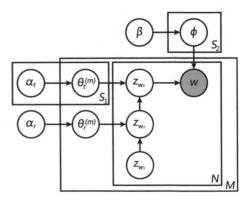

Algorithm 6: The generative process of a four-layer PAM

---

1 **for** *each sub-topic $k \in \{1, 2, \ldots, S_2\}$* **do**
2   generate a sub-topic-word distribution $\phi_k \sim$ Dirichlet(β);
3 **end**
4 **for** *each document d_m, $m \in \{1, 2, \ldots, M\}$* **do**
5   generate a super-topic distribution $\theta_r^{(m)} \sim$ Dirichlet(α_r);
6   generate a sub-topic distribution $\theta_t^{(m)} \sim$ Dirichlet(α_t);
7   **for** *each position in d_m $n \in \{1, 2, \ldots, N_m\}$* **do**
8   generate the topic path whose length is 3, using $L_w = < z_{w1}, z_{w2}, z_{w3} >$;
9   generate a word $w_{m,n} \sim$ Multinomial($\phi_{z_{w3}}$);
10   **end**
11 **end**

---

generate the topic path, and the sub-topics share the same format as the topics in LDA. The joint probability of a word w and topics z, θ_r^m, θ_t^m, and Φ in d_m is as follows:

$$p(w, z, \theta_r^m, \theta_t^m, \Phi | \alpha, \beta) =$$

$$p(\theta_r^{(m)} | \alpha_r) \prod_{i=1}^{S1} p(\theta_{t_i}^{(m)} | \alpha_{t_i}) \prod_w (p(z_{w2} | \theta_r^{(m)}) p(z_{w3} | \theta_{t_{z_{w2}}}^{(m)}) p(w | \phi_{z_{w3}})) \tag{2.11}$$

Figure 2.14 shows examples of topics generated by a PAM. A PAM typically provides superior representation to LDA, given that its hierarchical structure is more natural and contains richer information than a single-layer topic structure.

2.8.2 Rephil

Google's Rephil is another popular hierarchical topic model. Rephil is applied to Google AdSense [12], and substantially improves its performance. Figure 2.15 shows the graphical model of Rephil. Rephil can be regarded as a very complex Bayesian network containing 12 million word nodes, 1 million topic nodes, and 350 million edges [12]. Rephil and PAM share some common attributes: all leaf nodes are the words in the vocabulary, all latent nodes represent the topics, and different layers represent different levels of semantic abstraction. Their difference is that Rephil contains cross-layer edges and has more flexibility to model semantic relations.

Rephil assumes all variables are binary random variables: fired (indicated by 1) or unfired (indicated by 0). Rephil uses the noisy-OR as a conditional probability distribution between nodes. With noisy-OR, the effect of each parent node imposed on the child node is independent of the other parent node. This feature of noisy-OR simplifies the training process of Rephil, allowing it to be trained efficiently

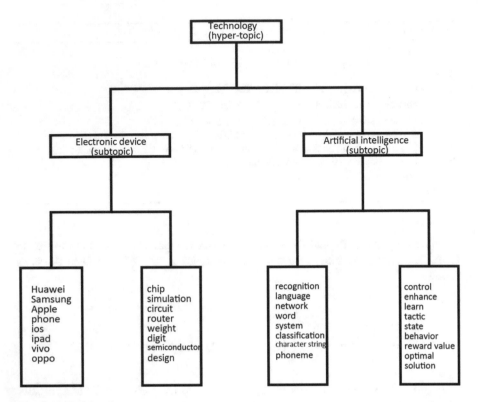

Fig. 2.14 PAM topics

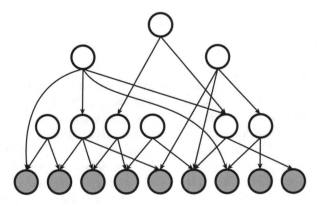

Fig. 2.15 Graphical model of Rephil

Fig. 2.16 An example of
noisy-OR

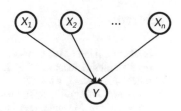

on massive text data. Figure 2.16 shows an example of the noisy-OR model. Each
parent node X_i can independently cause Y fired with a probability p_i. We can easily
obtain the probability that Y is fired by a given node X_i:

$$p_i = P(Y = 1|\bar{X}_1, \bar{X}_2, \dots, X_i, \dots, \bar{X}_n) \qquad (2.12)$$

We can further obtain the probability that Y is fired by a set of nodes X_s:

$$P(Y = 1|X_s) = 1 - \prod_{x_i \in X_s} (1 - p_i)^{x_i} \qquad (2.13)$$

In Rephil, a variant known as leaky noisy-OR is applied. The child node Y can
be fired even if none of the parent nodes is fired. This design is suitable for handling
situations in which Y can be fired by the other factors. Leaky noisy-OR introduces
a new concept, the leaky probability p_0, which is defined as follows:

$$p_0 = P(y|\bar{X}_1, \bar{X}_2, \dots, \bar{X}_i, \dots, \bar{X}_n), \qquad (2.14)$$

where p_0 represents a probability that a node Y will be fired if none of the parent
nodes X is fired.

Noisy-OR makes the parameters of Rephil to be linear in the number of parent
nodes, and greatly reduces the number of parameters that need to learn. However,
training Rephil remains a challenging task. Due to the need to learn both network
structure and parameters, Rephils is more difficult to train than topic models with
predefined structures. Meanwhile, the noisy-OR conditional probability distribution
is more suitable for modeling long-tail distributions of words, making Rephil able
to discover millions of long-tailed topics from massive amounts of data. When
applying Rephil to analyze the latent topics in a text, Rephil sets the value of the
leaf node of each observed word to 1 and calculates the posterior of the topic nodes.
The posterior of the topic nodes represents the latent semantics of the text.

References

1. Balikas G, Amini MR, Clausel M (2016) On a topic model for sentences. In: Proceedings of the 39th International ACM SIGIR Conference on Research and Development in Information Retrieval. ACM, pp 921–924
2. Blei DM, Lafferty JD (2006) Dynamic topic models. In: Proceedings of the 23rd International Conference on Machine Learning. ACM, pp 113–120
3. Blei DM, Ng AY, Jordan MI (2003) Latent Dirichlet allocation. J Mach Learn Res 3(Jan):993–1022
4. Deerwester S (1988) Improving information retrieval with latent semantic indexing. BibSonomy
5. Hofmann T (1999) Probabilistic latent semantic analysis. In: Proceedings of the Fifteenth Conference on Uncertainty in Artificial Intelligence. Morgan Kaufmann Publishers, Burlington, pp 289–296
6. Jiang D, Ng W (2013) Mining web search topics with diverse spatiotemporal patterns. In: Proceedings of the 36th International ACM SIGIR Conference on Research and Development in Information Retrieval, pp 881–884
7. Jiang D, Vosecky J, Leung KWT, Ng W (2012) G-wstd: A framework for geographic web search topic discovery. In: Proceedings of the 21st ACM International Conference on Information and Knowledge Management, pp 1143–1152
8. Jo Y, Oh AH (2011) Aspect and sentiment unification model for online review analysis. In: Proceedings of the Fourth ACM International Conference on Web Search and Data Mining. ACM, pp 815–824
9. Li W, McCallum A (2006) Pachinko allocation: Dag-structured mixture models of topic correlations. In: Proceedings of the 23rd International Conference on Machine Learning. ACM, pp 577–584
10. Liu Y, Liu Z, Chua TS, Sun M (2015) Topical word embeddings. In: Twenty-Ninth AAAI Conference on Artificial Intelligence, pp 2418–2424
11. Mikolov T, Chen K, Corrado G, Dean J (2013) Efficient estimation of word representations in vector space. Computer Science. arXiv:1301.3781
12. Murphy KP (2012) Machine learning: a probabilistic perspective. MIT Press, Cambridge
13. Sizov S (2010) Geofolk: latent spatial semantics in web 2.0 social media. In: Proceedings of the Third ACM International Conference on Web Search and Data Mining. ACM, pp 281–290
14. Wang X, Mccallum A (2006) Topics over time: a non-Markov continuous-time model of topical trends. In: ACM SIGKDD International Conference on Knowledge Discovery and Data Mining, pp 424–433

Chapter 3
Pre-processing of Training Data

Before training topic models, we must ensure the training data are appropriately formatted. Although existing literature rarely discusses pre-processing of training data, pre-processing plays a vital role in practice. Hence, this chapter details the tools and strategies required for pre-processing.

3.1 Word Segmentation

Space is used as the separator between words in English, making it easy to use English words as the "words" in topic models. However, no explicit separator between words and word segmentation is needed for languages such as Chinese. For example, the sentence "今天天气很好，我们出去外面放风筝吧。" is segmented into "今天__天气__很__好__，__我们__出去__外面__放__风筝__吧__。".

3.1.1 Chinese Word Segmentation Tools

Stanford CoreNLP [2] contains a set of language processing tools developed by Stanford University. Its word segmentation tool relies on the conditional random field and a large number of linguistic features such as character, morphology, repetition, etc. In addition to word segmentation, it has other functions such as part-of-speech tagging, named entity recognition, coreference resolution, and sentiment analysis.

NLPIR-Parser [4] is a language processing system developed by the Chinese Academy of Sciences. This system is compatible with mainstream operating systems and supports automatic recognition of out-of-vocabulary words as well as importing customized vocabulary. In addition, NLPIR-Parser contains common

Fig. 3.1 Vocabulary sizes of
different granularity

Document number	Fine-grained	Coarse-grained
100 thousand	1,044,430	1,780,729
1 million	4,987,771	7,927,704

functions used in natural language processing, such as part-of-speech tagging and named entity recognition.

Jieba[1] is a Chinese word segmentation component based on Python. Jieba supports customized word segmentation vocabulary, which can flexibly meet the personalized needs of users. Jieba also provides common functions, such as keyword extraction and part-of-speech tagging.

IKAnalyzer[2] is a Java-based word segmentation component. It supports multiple-word segmentation modes for different needs. It additionally facilitates the expansion of existing vocabularies.

3.1.2   Word Segmentation Granularity

Granularity is crucial for the result of word segmentation. For example, in coarse-grained word segmentation, "中华人民共和国" is regarded as a single word. In contrast, in fine-grained word segmentation, "中华人民共和国" is segmented into four words: "中华", "人民", "共和" and "国". Therefore, word segmentation granularity can significantly affect the vocabulary and segmentation results. Fig. 3.1 shows the vocabulary size obtained by different granularity. The vocabulary obtained by fine-grained word segmentation is significantly smaller than that of coarse-grained word segmentation. Specific application scenarios require a suitable word segmentation granularity. Coarse-grained word segmentation is selected if a single word requires clear semantic information. In contrast, fine-grained word segmentation is chosen if the vocabulary requires expansion to prevent the out-of-vocabulary issue.

3.2   Normalization

There are different expressions of the same word in the training corpus. To train a coherent topic model, we need to normalize different expressions. Normalization can be categorized into the following two types:

---

[1] https://github.com/fxsjy/jieba.

[2] https://github.com/wks/ik-analyzer.

- English case normalization: The same word may have different cases, such as "NBA" and "nba." Without case normalization, English words of different cases are treated as different words. Therefore, the English words are converted into uppercase or lowercase during pre-processing to alleviate this issue.
- Traditional/simplified Chinese normalization: Chinese texts can be written in traditional or simplified Chinese characters. These texts can be converted into the other format through a mapping vocabulary. Therefore, during pre-processing, the characters of the training texts are converted into either traditional or simplified versions.

3.3 Filtering

The training corpus may contain information that is useless for training. We filter such information in the pre-processing.

3.3.1 Stopword Filtering

Stopwords refer to punctuation and auxiliary words that offer little semantic information but frequently appear in documents. Their presence in the training data negatively impacts model training and topic interpretation. Filtering stopwords usually improve the training efficiency and the interpretability of the discovered topics. However, filtering stopwords is a double-edged sword. Excessive filtering increases the chance of encountering out-of-vocabulary words when inferring topics from new data. Therefore, practitioners must carefully determine an appropriate strength for filtering stopwords for specific tasks.

3.3.2 Low-Frequency Word Filtering

Statistical analysis shows that nearly half of the words within the vocabulary only appear once in the corpus. This low frequency makes it challenging for topic models to learn their latent semantics. Moreover, excessive low-frequency words result in an overly large vocabulary, which consumes substantial memory space during training.

The most straightforward approach to filtering low-frequency words involves establishing a threshold and filtering all the words whose frequency is less than the threshold. However, despite its simplicity, this strategy is plagued with some shortcomings. For example, if a small threshold is set on a massive data set, the vocabulary will still contain extensive words after filtering. On the contrary, if a relatively large threshold is set on a small dataset, the vocabulary will be overly shrunken and there will be frequent occurrences of out-of-vocabulary words. To solve this problem, vocabulary coverage is used to determine an optimal filtering threshold in practice. We count the frequencies of all the words in the training corpus and then choose the threshold according to the proportion of words to be preserved. Empirical analysis shows that a frequency threshold of keeping 99.5% of the words in the original vocabulary ensures coverage and effectively eliminates long-tail words.

3.3.3 Part-of-Speech-Based Filtering

Some words with certain part-of-speech tags (e.g., adverbs, conjunctions, prepositions, pronouns, and numerals) contain little semantic information. Therefore, filtering these types of words reduces the size of the vocabulary and improves training efficiency.

3.4 Word Sorting

Another important topic in data pre-processing is whether to maintain the original word order of the documents in training data. Some topic models (e.g., PLSA and LDA) adopt the bag-of-words assumption, whereby the original word order is disregarded, and a certain extent of efficiency advantage may be attained during training. For example, a distributed training framework of LDA named LightLDA [3] sorts the words in the documents of the training corpus. Such data format effectively reduces the information transmitted in the distributed network and significantly improves training efficiency.

In order to better model the contextual information of words, some topic models rely on the word order of the original document. For example, SentenceLDA [1] assumes that all the words in a sentence are generated from the same topic, and the sentence structure needs to be preserved in the training data. Similarly, the TWE model requires the original word order to be maintained. Since training the word vector and topic vector needs the contextual information of the target word, the word order cannot be disrupted.

References

1. Balikas G, Amini MR, Clausel M (2016) On a topic model for sentences. In: Proceedings of the 39th International ACM SIGIR Conference on Research and Development in Information Retrieval. ACM, New York, pp 921–924
2. Manning CD, Surdeanu M, Bauer J, Finkel JR, Bethard S, McClosky D (2014) The Stanford CoreNLP natural language processing toolkit. In: Proceedings of 52nd Annual Meeting of the Association for Computational Linguistics: System Demonstrations, pp 55–60

3. Yuan J, Gao F, Ho Q, Dai W, Wei J, Zheng X, Xing EP, Liu TY, Ma WY (2015) LightLDA: big topic models on modest computer clusters. In: Proceedings of the 24th International Conference on World Wide Web, International World Wide Web Conferences Steering Committee, pp 1351–1361

4. Zhang H, Shang J (2019) NLPIR-Parser: an intelligent semantic analysis toolkit for big data. In: Corpus Linguistics, vol 6, pp 87–104

Chapter 4
Expectation Maximization

The existence of latent variables increases the difficulty of estimating model parameters. The expectation maximization (EM) algorithm is a common method used to obtain the local optima of these parameters of probabilistic graphical models with latent variables. This chapter introduces how to apply the EM algorithm in topic models such as PLSA. The foundation of EM algorithm is introduced in Sect. 4.1. The convergence of EM is introduced in Sect. 4.2 and the generalized expectation maximization (GEM) is discussed in Sect. 4.3. Finally, the applications of EM and GEM in topic models are explained in Sect. 4.4.

4.1 Basics

The EM algorithm was originally proposed by Dempster et al. in 1977 [1]. Given x as a set of observed variables, z as a set of latent variables, and Θ as the model parameters, the maximum likelihood estimation of Θ is:

$$\mathcal{L}(\Theta) = \log p(x|\Theta) = \log \int p(x, z|\Theta)dz \tag{4.1}$$

However, due to the existence of the latent variables z, it is challenging to calculate the integral. The computational cost increases exponentially with the cardinality of z, and the EM algorithm can effectively solve this problem. In particular, the lower bound of $\mathcal{L}(\Theta)$ is firstly constructed in the E-step. Then, this lower bound is optimized in the M-step. This process alternates between these two steps and finally it reaches a local optima of Θ. Two typical methods in literature are used to deduce the lower bound of $\mathcal{L}(\Theta)$ in the E-step. In the following section, we explain the detailed differences and connections between these two methods.

© The Author(s), under exclusive license to Springer Nature Singapore Pte Ltd. 2023
D. Jiang et al., *Probabilistic Topic Models*,
https://doi.org/10.1007/978-981-99-2431-8_4

4.1.1   The First Method of E-Step

The parameter Θ is randomly initialized as $\Theta^{(0)}$. Consider $\Theta^{(t)}$ as the parameter value in the t-th iteration and suppose we have the distribution $p(z|x, \Theta^{(t)})$, given

- $\log(\cdot)$ is a concave function.
- $p(z|x, \Theta^{(t)}) \geq 0$.
- $\sum_z p(z|x, \Theta^{(t)}) = 1$.

we can obtain the lower bound of $\mathcal{L}(\Theta)$ using the Jensen's inequality:

$$\mathcal{L}(\Theta) = \log \sum_z p(z|x, \Theta^{(t)}) \frac{p(x, z|\Theta)}{p(z|x, \Theta^{(t)})} \tag{4.2a}$$

$$\geq \sum_z p(z|x, \Theta^{(t)}) \log \frac{p(x, z|\Theta)}{p(z|x, \Theta^{(t)})} \tag{4.2b}$$

$$= \sum_z p(z|x, \Theta^{(t)}) \log p(x, z|\Theta) - \sum_z p(z|x, \Theta^{(t)}) \log p(z|x, \Theta^{(t)}) \tag{4.2c}$$

The purpose of the EM algorithm is to continuously maximize this lower bound of $\mathcal{L}(\Theta)$. Given that the last term in Eq. (4.2c) is irrelevant to Θ, this term can be regarded as a constant. Hence, only the first term needs to be optimized:

$$E_{[z|x, \Theta^{(t)}]}[\log p(x, z|\Theta)] = \sum_z p(z|x, \Theta^{(t)}) \log p(x, z|\Theta) \tag{4.3}$$

The above equation is referred to as the **Q function**, denoted by $Q(\Theta, \Theta^{(t)})$.

4.1.2   The Second Method of E-Step

In the second method, the parameter Θ is still randomly initialized as $\Theta^{(0)}$ and $\Theta^{(t)}$ refers to the parameter value in the t-th iteration. Given $\mathcal{L}(\Theta) = \log p(x|\Theta) = \log p(x, z|\Theta) - \log p(z|x, \Theta)$, we have

$$\mathcal{L}(\Theta) - \mathcal{L}(\Theta^{(t)}) = \log p(x, z|\Theta) - \log p(x, z|\Theta^{(t)}) + \log \frac{p(z|x, \Theta^{(t)})}{p(z|x, \Theta)} \tag{4.4}$$

To find the expectation of $p(z|\boldsymbol{x}, \Theta^{(t)})$ for both sides of the above equation:

$$\mathcal{L}(\Theta) - \mathcal{L}(\Theta^{(t)}) = \sum_z p(z|\boldsymbol{x}, \Theta^{(t)}) \log p(\boldsymbol{x}, z|\Theta)$$

$$- \sum_z p(z|\boldsymbol{x}, \Theta^{(t)}) \log p(\boldsymbol{x}, z|\Theta^{(t)}) \tag{4.5}$$

$$+ \sum_z p(z|\boldsymbol{x}, \Theta^{(t)}) \log \frac{p(z|\boldsymbol{x}, \Theta^{(t)})}{p(z|\boldsymbol{x}, \Theta)}$$

The right side $\sum_z p(z|\boldsymbol{x}, \Theta^{(t)}) \log \frac{p(z|\boldsymbol{x},\Theta^{(t)})}{p(z|\boldsymbol{x},\Theta)}$ can be regarded as the KL divergences between $p(z|\boldsymbol{x}, \Theta^{(t)})$ and $p(z|\boldsymbol{x}, \Theta)$, namely $D_{KL}(p(z|\boldsymbol{x}, \Theta^{(t)})||p(z|\boldsymbol{x}, \Theta))$. Given that the value of the KL divergence is not negative, we have

$$\mathcal{L}(\Theta) \geq \mathcal{L}(\Theta^{(t)}) + \sum_z p(z|\boldsymbol{x}, \Theta^{(t)}) \log p(\boldsymbol{x}, z|\Theta)$$

$$- \sum_z p(z|\boldsymbol{x}, \Theta^{(t)}) \log p(\boldsymbol{x}, z|\Theta^{(t)}) \tag{4.6}$$

Then, we can obtain a lower bound of $\mathcal{L}(\Theta)$. Given that the first and the third terms of the right side in Eq. (4.6) are irrelevant to Θ, they are regarded as the constant terms. The second term (i.e., Eq. (4.3)) needs to be optimized and is called as the **Q function**.

4.1.3 M-Step

In the M-step, we aim to find $\Theta^{(t+1)}$ that maximizes the lower bound of $\mathcal{L}(\Theta)$, that is:

$$\Theta^{(t+1)} = \arg\max_{\Theta} Q(\Theta, \Theta^{(t)}) \tag{4.7}$$

Algorithm 1 summarizes the overall process of the EM algorithm. It alternates between the E-step and the M-step until convergence. The criterion of the convergence is that the parameters obtained by two consecutive iterations are not significantly updated (i.e., $\|\Theta^{(t+1)} - \Theta^{(t)}\| < \xi_1$, ξ_1 is a threshold), or the Q function by two consecutive iterations are not significantly updated (i.e., $\|Q(\Theta^{(t+1)}, \Theta^{(t)}) - Q(\Theta^{(t)}, \Theta^{(t)})\| < \xi_2$, ξ_2 is a threshold).

Algorithm 1: The EM algorithm

---

1 Input:an observed variable x, latent variable data z, the joint distribution $p(x, z|\Theta)$, the
   conditional distribution $p(z|x, \Theta)$
2 Output:a model parameter Θ
3 The initialized parameter is $\Theta^{(0)}$;
4 while *the convergence condition is not reached* **do**
5 | E-step:
6 | According to Equation (4.3), calculate $Q(\Theta, \Theta^{(t)})$;
7 | M-step:
8 | According to Equation (4.7), calculate $\Theta^{(t+1)}$;
9 end

4.2   Convergence of the EM Algorithm

In this section, we discuss the convergence of the EM algorithm. Let $\Theta = \Theta^{(t+1)}$ in Eq. (4.5), we have

$$
\mathcal{L}(\Theta^{(t+1)}) - \mathcal{L}(\Theta^{(t)})
$$

$$
= \sum_{z} p(z|x, \Theta^{(t)}) \log p(x, z|\Theta^{(t+1)}) - \sum_{z} p(z|x, \Theta^{(t)}) \log p(x, z|\Theta^{(t)})
$$

$$
+ \sum_{z} p(z|x, \Theta^{(t)}) \log \frac{p(z|x, \Theta^{(t)})}{p(z|x, \Theta^{(t+1)})}
$$

$$
= Q(\Theta^{(t+1)}, \Theta^{(t)}) - Q(\Theta^{(t)}, \Theta^{(t)}) + D_{KL}(p(z|x, \Theta^{(t)})||p(z|x, \Theta^{(t+1)}))
$$

$$
\tag{4.8}
$$

Given that $\Theta^{(t+1)}$ makes $Q(\Theta, \Theta^{(t)})$ reach the maximum:

$$
Q(\Theta^{(t+1)}, \Theta^{(t)}) - Q(\Theta^{(t)}, \Theta^{(t)}) \geq 0 \tag{4.9}
$$

Given that the value of KL divergence is always not negative:

$$
D_{KL}(p(z|x, \Theta^{(t)})||p(z|x, \Theta^{(t+1)})) \geq 0 \tag{4.10}
$$

Based upon Eqs. (4.9) and (4.10), we get $\mathcal{L}(\Theta^{(t+1)}) \geq \mathcal{L}(\Theta^{(t)})$, indicating that this likelihood is monotonically increasing. The EM algorithm is not guaranteed to converge to the global optima. Different initial value may result in different local optima. In practice, we can try different initial values to improve the quality of the results.

4.3 GEM Algorithm

The generalized expectation maximization (GEM) algorithm is a variant of the EM algorithm. Sometimes it is challenging to obtain the set of parameters that maximizes $Q(\Theta, \Theta^{(t)})$ in the M-step. The GEM algorithm is proposed to solve this problem by only requiring a set of parameters that improve $Q(\Theta, \Theta^{(t)})$ rather than the best set that maximizes $Q(\Theta, \Theta^{(t)})$. As shown in Algorithm 2, GEM only requires that $\Theta^{(t+1)}$ in the M-step satisfies $Q(\Theta^{(t+1)}, \Theta^{(t)}) \geq Q(\Theta^{(t)}, \Theta^{(t)})$. Meeting the above condition is sufficient to ensure $\mathcal{L}(\Theta^{(t+1)}) \geq \mathcal{L}(\Theta^{(t)})$ during each iteration. Since the set of parameter $\Theta^{(t+1)}$ in each iteration is not optimal, the convergence speed of the GEM algorithm is usually slower than that of the EM algorithm. Readers can refer to the relevant literature [3] to learn more details about GEM.

Algorithm 2: The GEM algorithm

---

1 Input: an observed variable x; a latent data z; the joint distribution $p(x, z|\Theta)$; the conditional distribution $p(z|x, \Theta)$
2 Output: a model parameter Θ
3 The initialized parameter is $\Theta^{(0)}$;
4 **while** *the convergence condition is not reached* **do**
5   E-step:
6   According to Equation (4.3), calculate $Q(\Theta|\Theta^{(t)})$;
7   M-step:
8   Calculate $\Theta^{(t+1)}$ to satisfy $Q(\Theta^{(t+1)}, \Theta^{(t)}) \geq Q(\Theta^{(t)}, \Theta^{(t)})$;
9 **end**

4.4 Applications of the EM Algorithm

4.4.1 PLSA

We use the probabilistic latent semantic analysis (PLSA) model [2] as an example to explain how to apply EM algorithm. Two multinomial distributions $p(z|d)$ and $p(w|z)$ constitute the PLSA parameters, i.e., $\Theta = \{p(w|z_j), p(z_j|d)|w \in \mathcal{V}, d \in \mathcal{D}, 1 \leq j \leq K\}$, where $\mathcal{D}$ is the corpus, $\mathcal{V}$ is the vocabulary, and K is the number of topics. Training PLSA model is to optimize Θ by maximizing the likelihood function of all documents in the corpus. The likelihood is as follows:

$$
\begin{aligned}
\mathcal{L}(\Theta) &= \log p(D|\Theta) \\
&= \sum_{d \in \mathcal{D}} \sum_{w \in \mathcal{V}} n(d, w) \log p(d, w) \\
&= \sum_{d \in \mathcal{D}} \left[n(d) \log p(d) + \sum_{w \in \mathcal{V}} n(d, w) \log \sum_{z \in \mathcal{Z}} p(w|z) p(z|d) \right]
\end{aligned}
\tag{4.11}
$$

where $n(d) = \sum_{w \in \mathcal{V}} n(d, w)$. $n(d, w)$ represents the frequency of a word w that appears in a document d. $n(d) \log p(d)$ is constant and can be ignored in the optimization process. According to the EM algorithm, we first need to write the Q function in the E-step:

$$
Q(\Theta, \Theta') = \sum_{d \in \mathcal{D}} \sum_{w \in \mathcal{V}} n(d, w) \sum_{z \in \mathcal{Z}} p(z|d, w) \log p(w|z) p(z|d)
\tag{4.12}
$$

According to the Bayes' theorem, $p(z|d, w)$ of the above equation is calculated as:

$$
p(z|d, w) = \frac{p(w|z) p(z|d)}{\sum_{z' \in \mathcal{Z}} p(w|z') p(z'|d)}
\tag{4.13}
$$

In the first iteration, $p(w|z)$ and $p(z|d)$ of the above equation are randomly initialized and they satisfy $\sum_{w \in \mathcal{V}} p(w|z) = 1$ and $\sum_{z \in \mathcal{Z}} p(z|d) = 1$. Then, $p(w|z)$ and $p(z|d)$ will be assigned to the values from the previous iteration of M-step. In the M-step, we maximize the Q function. Given the constraints $\sum_{w \in \mathcal{V}} p(w|z) = 1$ and $\sum_{z \in \mathcal{Z}} p(z|d) = 1$, the Lagrange function can be obtained by the Lagrange multiplier method:

$$
\mathcal{F} = Q(\Theta, \Theta') + \alpha \sum_{z \in \mathcal{Z}} \left(1 - \sum_{w \in \mathcal{V}} p(w|z) \right) + \beta \sum_{d \in \mathcal{D}} \left(1 - \sum_{z \in \mathcal{Z}} p(z|d) \right)
$$

By deriving $\mathcal{F}$, we can obtain the following equation:

$$
p(w|z) = \frac{\sum_{d \in \mathcal{D}} n(d, w) p(z|d, w)}{\sum_{w' \in \mathcal{V}} \sum_{d \in \mathcal{D}} n(d, w) p(z|d, w')}
\tag{4.14}
$$

$$
p(z|d) = \frac{\sum_{w \in \mathcal{V}} n(d, w) p(z|d, w)}{n(d)}
\tag{4.15}
$$

Through the above two equations, we can optimize the parameter Θ of PLSA.

Algorithm 3: Applying EM algorithm to PLSA

---

1 Initialize $P(w|z)$ and $P(z|d)$;
2 **while** *the convergence condition is not reached* **do**
3 | E-step:
4 | According to Equation (4.13), calculate $p(z|d, w)$;
5 | M-step:
6 | According to Equation (4.14) and Equation (4.15), calculate $p(w|z)$ and $p(z|d)$;
7 **end**

Algorithm 3 presents the EM algorithm solving process of PLSA. For PLSA, $p(z|d, w)$ is calculated in the E-step and $p(w|z)$ and $p(z|d)$ are calculated in the M-step. EM algorithm can not only be used to solve the original PLSA model but also some PLSA variants. The common feature of these models is that they are not complete Bayesian models and usually have explicit distinction between the parameters and the latent variables.

4.4.2   PCLSA

We use the probabilistic cross-lingual latent semantic analysis (PCLSA) model [4] as an example to explain how to apply GEM. Since PLSA models the co-occurrence of words in the document and different languages typically do not appear in the same document, the task of cross-lingual topic extraction is not supported by PLSA. PCLSA introduces a translation vocabulary to discover cross-lingual topics. As shown in Fig. 4.1, topics discovered by PCLSA contain multilingual words, and the words under the same topic have semantic correlations.

Taking a corpus containing Chinese and English documents as an example, Fig. 4.2 shows a bipartite graph of a Chinese-English translation vocabulary. A Chinese word corresponds to multiple English words (e.g., the Chinese word "河岸" can be translated into the English word "bank" or "strand"). Similarly, an English word corresponds to multiple Chinese words (e.g., the English word "bank" can be translated into the Chinese word "银行" or "河岸"). In this bipartite graph, an edge represents the translation relationship between the two words. The weight of the edge represents the probability that a word can be translated into the other. For PCLSA, the likelihood of a corpus is:

$$\mathcal{L} = \sum_{i=1}^{s} \sum_{d \in C_i} \sum_{w} n(w, d) \log \sum_{j=1}^{k} p(z_j|d) p(w|z_j) \tag{4.16}$$

where s represents the number of languages, C_i represents the corpus that only contains the i-th language, and $n(w, d)$ represents the word frequency of a word

Fig. 4.1 Examples of
cross-lingual topics

Topic 1	Topic 2	Topic 3
股市 (stock)	kill	korea
bank	iraq	合作 (collaboration)
投资者 (investor)	army	政府 (government)
market	伊拉克 (iraq)	会谈 (talks)
deal	炸弹 (bomb)	relation
invest	袭击 (attack)	meet
企业 (company)	iraqi	上海 (shanghai)
增长 (growth)	军队 (army)	country
基金 (fund)	police	两国 (bilateral)
trade	暴力 (violence)	会晤 (meet)

Fig. 4.2 A bipartite graph of
Chinese and English words

w within a document d. For the above example, $s = 2$, since we only have two
languages. Based on the likelihood in Eq. (4.16), PCLSA introduces a constraint:

$$\mathcal{R} = \frac{1}{2} \sum_{<u,v> \in E} w(u, v) \sum_{j=1}^{k} \left(\frac{p(w_u|z_j)}{Deg(u)} - \frac{p(w_v|z_j)}{Deg(v)} \right)^2 \qquad (4.17)$$

where $<u, v> \in E$ represents an edge connecting a Chinese word u and an English
word v, and $w(u, v)$ represents the weight of this edge $<u, v>$. $Deg(u)$ and
$Deg(v)$ refer to the out-degrees of node u and node v, respectively. Intuitively, $\mathcal{R}$
constrains that if two words have some translation relationship, the probabilities

that these two words appear in the same topic are higher. In summary, the objective function O that PCLSA maximizes is as follows:

$$O = (1 - \lambda)\mathcal{L} - \lambda\mathcal{R} \tag{4.18}$$

where $\lambda \in (0, 1)$. PCLSA applies the GME algorithm to optimize the parameters. In the E-step, we have:

$$p(z_j|w, d) = \frac{p(z_j|d)p(w|z_j)}{\sum_{j'} p(z_{j'}|d)p(w|z_{j'})} \tag{4.19}$$

In the M-step, the objective is to optimize

$$Q(\Theta, \Theta^t) = (1 - \lambda)\mathcal{L}' - \lambda\mathcal{R} \tag{4.20}$$

where $\mathcal{L}' = \sum_d \sum_w c(w, d) \sum_j z(w, d, j) \log p(z_j|d)p(w|z_j)$, $\sum_j \log p(z_j|d) = 1$ and $\sum_w \log p(w|z_j) = 1$. Like PLSA, PCLSA has a closed-form solution by maximizing only the $\mathcal{L}'$ part in Eq. (4.20):

$$p(w|z_j) = \frac{\sum_d n(w, d)p(z_j|w, d)}{\sum_d \sum_{w'} n(w', d)p(z_j|w', d)} \tag{4.21}$$

$$p(z_j|d) = \frac{\sum_w n(w, d)p(z_j|w, d)}{\sum_w \sum_{j'} n(w, d)p(z_{j'}|w, d)} \tag{4.22}$$

However, there is no closed-form solution when the term $\mathcal{R}$ is added. According to the GEM, there is no need to obtain the optimal set of parameters in each M-step. It only requires a new set of parameters that can improve the regularized likelihood function (i.e., $Q(\Theta^{(t+1)}, \Theta^{(t)}) > Q(\Theta^{(t)}, \Theta^{(t)})$). The GEM algorithm first maximizes $\mathcal{L}'$ according to Eq. (4.21) and then gradually optimizes $\mathcal{R}$ according to the following equation:

$$p^{t+1}(w_u|z_j) = (1 - \alpha)p^t(w_u|z_j) + \alpha \sum_{<u,v>\in E} \frac{w(u, v)}{Deg(v)} p^t(w_v|z_j) \tag{4.23}$$

where α is a weight parameter. The initial $p^t(w_u|z_j)$ can be obtained from Eq. (4.21). Then, Eq. (4.23) is repeatedly applied until the condition $Q(\Theta^{(t+1)}, \Theta^{(t)}) > Q(\Theta^{(t)}, \Theta^{(t)})$ cannot be satisfied and the parameter set Θ^t is the local optima. As shown in Algorithm 4, the process of the GEM algorithm is similar to the EM algorithm. The primary difference between these two algorithms is the M-step. Since GEM could be used to optimize the models when there is no closed-form solution, it has much wider applicability than the original EM algorithm.

Algorithm 4: Applying the GEM algorithm to PCLSA

---

1 while *the convergence condition is not reached* **do**
2   E-step:
3   According to Equation (4.19), calculate $p(z|d, w)$;
4   M-step:
5   According to Equation (4.21), calculate $p(w|z)$; according to Equation (4.22), calculate $p(z|d)$;
6   **while** O *still increases* **do**
7   According to Equation (4.23), update $p(w|z)$;
8   **end**
9 end

References

1. Dempster AP, Laird NM, Rubin DB (1977) Maximum likelihood from incomplete data via the EM algorithm. J R Stat Soc Ser B (Methodol) 39(1):1–22
2. Hofmann T (1999) Probabilistic latent semantic indexing. In: Proceedings of the 22nd Annual International ACM SIGIR Conference on Research and Development in Information Retrieval, pp 50–57
3. McLachlan GJ, Krishnan T (2007) The EM algorithm and extensions, vol 382. John Wiley & Sons, Hoboken
4. Zhang D, Mei Q, Zhai C (2010) Cross-lingual latent topic extraction. In: Proceedings of the 48th Annual Meeting of the Association for Computational Linguistics, pp 1128–1137

Chapter 5
Markov Chain Monte Carlo Sampling

The EM algorithm is suitable for topic models whose parameters and latent variables are distinguished. However, the approximate inference is usually applied in practice for fully Bayesian models whose parameters are treated as random variables. This chapter discusses an essential type of approximate inference: Markov chain Monte Carlo (MCMC) sampling.

5.1 Markov Chain

Markov process is a random process that exhibits Markov property, i.e., the probability of the successive state is determined only by the current state, with no dependence on the past or future states. Markov chain is a Markov process with discrete time and states. The probability of state change in a Markov chain is known as the transition probability. The matrix used to represent the transition probabilities between all states is known as the transition matrix and is usually denoted by Q. After the nth transition, the state distribution of a Markov chain with the initial state π_0 is calculated as follows:

$$\pi_n(x) = \pi_{n-1}(x)Q = \pi_0(x)Q^n \tag{5.1}$$

If the distribution $\pi_n(x)$ satisfies $\pi_n(x) = \pi_n(x)Q$, $\pi_n(x)$ is called the stationary distribution of the Markov chain.

To determine whether a Markov chain is able to converge, the concept of ergodicity is introduced, and it requires that the state space of the Markov chain is finite, i.e., any state can be reached from any other state via transition.

D. Jiang et al., *Probabilistic Topic Models*,
https://doi.org/10.1007/978-981-99-2431-8_5

Mathematically, if $Q(i, j)$ denotes the transition probability from the state i to the state j, the Markov chain will always converge to a stationary distribution $\pi(x)$, if

$$\pi(i)Q(i, j) = \pi(j)Q(j, i) \tag{5.2}$$

Equation (5.2) is also known as the detailed balance condition. With any initial state passing sufficient transitions, the samples of an ergodic Markov chain will follow the stationary distribution.

5.2   Monte Carlo Method

The Monte Carlo method transforms the target problem into a random distribution of values and evaluates the numerical properties of the target problem by sampling from the random distribution. The Monte Carlo method is often used to solve complex multi-dimensional integration, which is difficult to solve analytically. The Monte Carlo method can be used to obtain an approximate value. For example, consider the integration problem shown in Fig. 5.1:

$$s = \int_a^b f(x)dx \tag{5.3}$$

The integral s can be approximated by taking n samples $(x_1, \ldots, x_n)$ from the probability distribution $p(x)$ over the range $[a, b]$, and applying the following equation:

$$\int_a^b \frac{f(x)}{p(x)} p(x)dx \approx \frac{1}{n} \sum_{i=1}^n \frac{f(x_i)}{p(x_i)} = \hat{s}_n \tag{5.4}$$

According to the law of large numbers, if the samples are independent and identically distributed, their average value will converge to the expected value:

$$\lim_{n \to \infty} \hat{s}_n = s \tag{5.5}$$

The effectiveness of this approach depends on the convenience of sampling from the distribution $p(x)$. When sampling from $p(x)$ is difficult due to its complex form or high dimensionality, we resort to Markov chain Monte Carlo.

Fig. 5.1 Integration of $f(x)$ over $[a, b]$

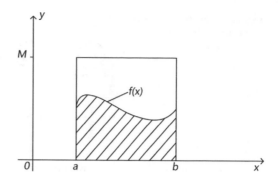

5.3 Markov Chain Monte Carlo

By constructing a Markov chain with a transition matrix Q and a stationary distribution $p(x)$, the samples taken after the chain converges can be regarded as samples from $p(x)$. The period before the Markov chain reaches convergence is known as the burn-in period. The samples obtained during the burn-in period are not considered as the samples from the target distribution and should be discarded.

For the target distribution $p(x)$, a Markov chain does not necessarily satisfy the detailed balance condition:

$$p(i)Q(i, j) \neq p(j)Q(j, i) \tag{5.6}$$

Both the left and right sides of Eq. (5.6) are multiplied by a coefficient $\alpha(i, j)$ to ensure that the detailed balance condition holds

$$p(i)Q(i, j)\alpha(i, j) = p(j)Q(j, i)\alpha(j, i) \tag{5.7}$$

A proper coefficient $\alpha(i, j)$ can be constructed as follows:

$$\alpha(i, j) = p(j)Q(j, i), \alpha(j, i) = p(i)Q(i, j) \tag{5.8}$$

We further obtain

$$p(i)\underbrace{Q(i, j)\alpha(i, j)}_{Q'(i,j)} = p(j)\underbrace{Q(j, i)\alpha(j, i)}_{Q'(j,i)} \tag{5.9}$$

Based upon the above modifications, the Markov chain with a transition matrix Q becomes one with a transition matrix Q' and the stationary distribution $p(x)$. The coefficient $\alpha(i, j)$ is known as the acceptance ratio: as we jump from a state i to a state j with the probability of $Q(i, j)$ in the original Markov chain, we accept the transition with the probability of $\alpha(i, j)$ and thus obtain the transition probability of the new Markov chain $Q' = Q(i, j)\alpha(i, j)$. This process is depicted in Fig. 5.2.

Fig. 5.2 State transition and acceptance probability of a Markov chain

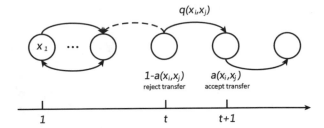

Algorithm 1 presents the overall process of MCMC. The algorithm samples the next state x according to the $p(x|x_t)$ distribution at time t. A random number is then sampled from the uniform distribution $[0, 1]$ and compared with the acceptance rate $\alpha(x_t, x)$ to decide whether to accept the state transition. This process is repeated until sufficient samples are generated. MCMC covers a wide variety of specific sampling algorithms, and the ones commonly used for topic models are Gibbs sampling and Metropolis–Hastings sampling.

Algorithm 1: Markov chain Monte Carlo

---

1 randomly set the initial state x_0
2 **for** $t = 0,1,2,\ldots$ **do**
3 $\quad$ sample $x \sim p(x|x_t)$;
4 $\quad$ generate a random number u from the uniform distribution $[0, 1]$;
5 $\quad$ **if** $p_i < 1$ **then**
6 $\quad\quad$ accept the state transfer, $x_{t+1} = x$;
7 $\quad$ **else**
8 $\quad\quad$ deny the state transfer, $x_{t+1} = x_t$;
9 $\quad$ **end**
10 **end**

Two important factors should be considered when applying MCMC: the number of Markov chains and the number of samples. There are four existing strategies: single final (SF), single average (SA), multiple final (MF), and multiple average (MA) [2]:

- SF takes the final sample of a single Markov chain as the result.
- SA takes the average value of multiple samples of a single Markov chain as the result.
- MF takes the average value of the final samples of multiple Markov chains as the result.
- MA takes the average value of multiple samples of multiple Markov chains as the result.

Experiments show that the best performance is achieved by taking the average value of multiple sampling results of multiple Markov chains [2]. However, using multiple Markov chains has some inherent disadvantages, including the requirement for additional iterations during the burn-in period, and thus generates more invalid samples. For the sake of efficiency, in the field of topic models, the common practice is to adopt a single Markov chain and use the final sampling results only.

5.4 Gibbs Sampling

5.4.1 Basic Concepts

Gibbs sampling works by fixing the values of the other variables and sampling the value of the current variable x_i with a conditional distribution $p(x_i|\mathbf{x}_{-i})$, where $\mathbf{x}_{-i}$ denotes $\mathbf{x}$ removing the variable x_i. The sampling process proceeds iteratively until the Markov chain reaches a stationary distribution. Sampling from this stationary distribution is equivalent to sampling from the joint distribution $p(\mathbf{x})$.

Consider Gibbs sampling in three-dimensional space. As shown in Fig. 5.3, there exist two points $A(x_1, y_1, z_1)$ and $B(x_2, y_1, z_1)$. The y coordinates and z coordinates of both points are identical. As such, we can derive the following equation:

$$p(x_1, y_1, z_1)p(x_2|y_1, z_1) = p(y_1, z_1)p(x_1|y_1, z_1)p(x_2|y_1, z_1)$$
$$p(x_2, y_1, z_1)p(x_1|y_1, z_1) = p(y_1, z_1)p(x_2|y_1, z_1)p(x_1|y_1, z_1)$$

$$(5.10)$$

From this, we obtain

$$p(x_1, y_1, z_1)p(x_2|y_1, z_1) = p(x_2, y_1, z_1)p(x_1|y_1, z_1) \tag{5.11}$$

We further get

$$p(A)p(x_2|y_1, z_1) = p(B)p(x_1|y_1, z_1) \tag{5.12}$$

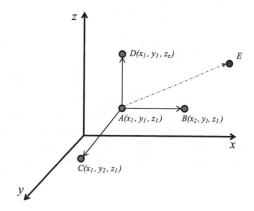

Fig. 5.3 The construction of a Markov chain matrix in three-dimensional space

The above equations demonstrate that along the line given by $y = y_1$, $z = z_1$, with a transition probability $p(x|y_1, z_1)$, the transition between any two points satisfies the detailed balance condition. Following the same procedure, two arbitrary points $A(x_1, y_1, z_1)$ and $C(x_1, y_2, z_1)$ along the line given by $x = x_1$, $z = z_1$ are related as follows:

$$p(A)p(y_2|x_1, z_1) = p(C)p(y_1|x_1, z_1) \tag{5.13}$$

Similarly, two arbitrary points $A(x_1, y_1, z_1)$ and $D(x_1, y_1, z_2)$ along the line given by $x = x_1$, $y = y_1$ are related as follows:

$$p(A)p(z_2|x_1, y_1) = p(D)p(z_1|x_1, y_1) \tag{5.14}$$

The above equations can be combined to construct a following state transition matrix Q:

$$\begin{cases} Q(A \to B) = p(x_B|y_1, z_1) & \text{if } y_A = y_B = y_1, z_A = z_B = z_1 \\ Q(A \to C) = p(y_B|x_1, z_1) & \text{if } x_A = x_B = x_1, z_A = z_B = z_1 \\ Q(A \to D) = p(z_B|x_1, y_1) & \text{if } x_A = y_B = x_1, y_A = y_B = y_1 \\ \quad Q(A \to E) = 0 & \text{else.} \end{cases} \tag{5.15}$$

Based upon Q, we can easily verify that the detailed balance condition is satisfied for any two points in three-dimensional space. Hence, the Markov chain in this space will converge to the stationary distribution $p(x)$.

For the joint distribution $p(x, y, z)$ in the above example, during each step, the Gibbs sampling algorithm updates the value of a given variable, while the other two variables are fixed. Assume that the values of these three variables are $x(\tau)$, $y(\tau)$, and $z(\tau)$ during the step τ of the algorithm. The step $\tau + 1$ proceeds as follows:

- Replace $x(\tau)$ with a new value $x(\tau + 1)$, which is sampled from the probability distribution $p(x|y(\tau), z(\tau))$.
- Replace $y(\tau)$ with a new value $y(\tau + 1)$, which is sampled from the probability distribution $p(y|x(\tau + 1), z(\tau))$.
- Replace $z(\tau)$ with a new value $z(\tau + 1)$, which is sampled from the probability distribution $p(z|x(\tau + 1), y(\tau + 1))$.

This process can be easily extended to higher dimensions. When $x = (x_1, \ldots, x_u)$, we obtain the general Gibbs sampling algorithm in Algorithm 2.

In practice, collapsed Gibbs sampling is used to improve sampling efficiency. Collapsed Gibbs sampling integrates out some variables and only samples the remaining variables. Due to the reduced dimensionality of the sampling space, collapsed Gibbs sampling is more efficient than the original Gibbs sampling.

Algorithm 2: Gibbs sampling

1 random initial state $\{x_j : j = 1, \ldots, u\}$
2 set $t = 0$
3 **for** $t = 0, 1, 2, \ldots$ **do**
4 $\quad$ $x_1^{(t+1)} \sim p(x_1|x_2^{(t)}, x_3^{(t)}, \ldots, x_u^{(t)})$
5 $\quad$ $x_2^{(t+1)} \sim p(x_2|x_1^{(t+1)}, x_3^{(t)}, \ldots, x_u^{(t)})$
6 $\quad$ $\ldots$
7 $\quad$ $x_j^{(t+1)} \sim p(x_j|x_1^{(t+1)}, \ldots, x_{j-1}^{(t+1)}, x_{j+1}^{(t)}, \ldots, x_u^{(t)})$
8 $\quad$ $\ldots$
9 $\quad$ $x_u^{(t+1)} \sim p(x_u|x_1^{(t)}, x_2^{(t)}, \ldots, x_{u-1}^{(t+1)})$
10 **end**

5.4.2 Application of Gibbs Sampling in LDA

The Gibbs sampling is particularly suitable for sampling the posterior of Bayesian networks. In this section, we use LDA to demonstrate the specific process of applying Gibbs sampling. In LDA, the joint probability distribution of w, z, θ_m, and Φ in LDA is

$$
p(w, z, \theta_m, \Phi|\alpha, \beta) = \underbrace{p(\Phi|\beta)}_{\text{Topic Plate}} p(\theta_m|\alpha) \overbrace{\prod_{n=1}^{N_m} p(w_{m,n}|\phi_{z_{m,n}}) p(z_{m,n}|\theta_m)}^{\text{Document Plate}} \tag{5.16}
$$

By integrating θ_m and Φ, we obtain the joint distribution of the words w and the topics z within a document d_m:

$$
p(w, z|\alpha, \beta) = \int \int p(\theta|\alpha)p(\phi|\beta) \prod_{n=1}^{N} \sum_{z_n} p(z_n|\theta) p(w_n|\phi_{z_n}) d\phi d\theta p(w, z|\alpha, \beta)
$$

$$
= \int \int p(w, z, \theta_m, \Phi|\alpha, \beta) \mathrm{d}\Phi \mathrm{d}\theta_m
$$

$$
= \int \int p(\theta_m|\alpha)p(\Phi|\beta) \prod_{n=1}^{N_m} p(w_{m,n}|\phi_{z_{m,n}}) p(z_{m,n}|\theta_m) \mathrm{d}\Phi \mathrm{d}\theta_m
$$

$$
= \int p(\theta_m|\alpha) \prod_{n=1}^{N_m} p(z_{m,n}|\theta_m) \mathrm{d}\theta_m \int p(\Phi|\beta) \prod_{n=1}^{N_m} p(w_{m,n}|\phi_{z_{m,n}}) \mathrm{d}\Phi
$$

$$
\tag{5.17}
$$

The integral of θ_m is given as follows:

$$\int p(\boldsymbol{\theta_m}|\alpha) \prod_{n=1}^{N_m} p(z_{m,n}|\boldsymbol{\theta_m})\mathrm{d}\boldsymbol{\theta_m} = \int \frac{\Gamma(\Sigma_{k=1}^{K}\alpha_k)}{\Pi_{k=1}^{K}\Gamma(\alpha_k)} \prod_{k=1}^{K} \theta_k^{\alpha_k-1} \prod_{n=1}^{N_m} p(z_{m,n}|\boldsymbol{\theta_m})\mathrm{d}\boldsymbol{\theta_m},$$
$$(5.18)$$

where $\prod_{n=1}^{N_m} p(z_{m,n}|\boldsymbol{\theta_m})$ in Eq. (5.18) can be expressed as

$$\prod_{n=1}^{N_m} p(z_{m,n}|\boldsymbol{\theta_m}) = \prod_{k=1}^{K} \theta_m^{n_m^k} \qquad (5.19)$$

Therefore, Eq. (5.18) can be rewritten as

$$\int \frac{\Gamma(\Sigma_{k=1}^{K}\alpha_k)}{\Pi_{k=1}^{K}\Gamma(\alpha_k)} \prod_{k=1}^{K} \theta_k^{\alpha_k-1} \prod_{k=1}^{K} \theta_m^{n_m^k}\mathrm{d}\boldsymbol{\theta_m} = \int \frac{\Gamma(\Sigma_{k=1}^{K}\alpha_k)}{\Pi_{k=1}^{K}\Gamma(\alpha_k)} \prod_{k=1}^{K} \theta_k^{n_m^k+\alpha_k-1}\mathrm{d}\boldsymbol{\theta_m}$$
$$(5.20)$$

The right side of Eq. (5.20) can be transformed as

$$\int p(\boldsymbol{\theta_m}|\alpha) \prod_{n=1}^{N_m} p(z_{m,n}|\boldsymbol{\theta_m})\mathrm{d}\boldsymbol{\theta_m}$$

$$= \int \frac{\Gamma(\Sigma_{k=1}^{K}\alpha_k)}{\Pi_{k=1}^{K}\Gamma(\alpha_k)} \prod_{k=1}^{K} \theta_k^{n_m^k+\alpha_k-1}\mathrm{d}\boldsymbol{\theta_m}$$

$$= \frac{\Gamma(\Sigma_{k=1}^{K}\alpha_k)}{\Pi_{k=1}^{K}\Gamma(\alpha_k)} \frac{\Pi_{k=1}^{K}\Gamma(n_m^k+\alpha_k)}{\Gamma(\Sigma_{k=1}^{K}n_m^k+\alpha_k)} \int \frac{\Gamma(\Sigma_{k=1}^{K}n_m^k+\alpha_k)}{\Pi_{k=1}^{K}\Gamma(n_m^k+\alpha_k)} \prod_{k=1}^{K} \theta_k^{n_m^k+\alpha_k-1}\mathrm{d}\boldsymbol{\theta_m}$$

$$= \frac{\Gamma(\Sigma_{k=1}^{K}\alpha_k)}{\Pi_{k=1}^{K}\Gamma(\alpha_k)} \frac{\Pi_{k=1}^{K}\Gamma(n_m^k+\alpha_k)}{\Gamma(\Sigma_{k=1}^{K}n_m^k+\alpha_k)}, \qquad (5.21)$$

where n_m^k represents the number of words associated with a topic k within the m-th document. Similarly, the integral of Φ is given as follows:

$$\int p(\Phi|\beta) \prod_{n=1}^{N_m} p(w_{m,n}|\phi_{z_{m,n}})\mathrm{d}\Phi = \prod_{k=1}^{K} \frac{\Gamma(\sum_{r=1}^{V}\beta_r)}{\Pi_{r=1}^{V}\Gamma(\beta_r)} \frac{\Pi_{r=1}^{V}\Gamma(n_r^k+\beta_r)}{\Gamma(\sum_{r=1}^{V}n_r^k+\beta_r)} \qquad (5.22)$$

Based on Eqs. (5.21) and (5.22), we obtain

$$p(\boldsymbol{w}, \boldsymbol{z} | \alpha, \beta)$$

$$= \prod_{k=1}^{K} \frac{\Gamma(\sum_{r=1}^{V} \beta_r) \prod_{r=1}^{V} \Gamma(n_r^k + \beta_r)}{\prod_{r=1}^{V} \Gamma(\beta_r) \ \Gamma(\sum_{r=1}^{V} n_r^k + \beta_r)} * \frac{\Gamma(\Sigma_{k=1}^{K} \alpha_k) \ \Pi_{k=1}^{K} \Gamma(n_m^k + \alpha_k)}{\Pi_{k=1}^{K} \Gamma(\alpha_k) \ \Gamma(\Sigma_{k=1}^{K} n_m^k + \alpha_k)}$$

$$(5.23)$$

Gibbs sampling of LDA needs the conditional probability of a topic for a word, and this probability is as follows:

$$p(z_{(m,n)}$$

$$= k | z_{-(m,n)}, \boldsymbol{w}, \alpha, \beta)$$

$$= \frac{p(\boldsymbol{w}, \boldsymbol{z} | \alpha, \beta)}{p(\boldsymbol{w}, \boldsymbol{z}_{-(\boldsymbol{m},\boldsymbol{n})} | \alpha, \beta)}$$

$$= \frac{p(\boldsymbol{w} | \boldsymbol{z}, \alpha, \beta)}{p(\boldsymbol{w}_{-(\boldsymbol{m},\boldsymbol{n})} | \boldsymbol{z}_{-(\boldsymbol{m},\boldsymbol{n})}, \alpha, \beta) p(w_{(m,n)}, \alpha, \beta)} \cdot \frac{p(\boldsymbol{z}, \alpha, \beta)}{p(\boldsymbol{z}_{-(\boldsymbol{m},\boldsymbol{n})}, \alpha, \beta)}$$

$$\propto \frac{\Gamma(n_k^t + \beta_t) \Gamma(\sum_{t=1}^{V} n_{k,-(m,n)}^t + \beta_t)}{\Gamma(n_{k,-(m,n)}^t + \beta_t) \Gamma(\sum_{t=1}^{V} n_k^t + \beta_t)} \cdot \frac{\Gamma(n_m^k + \alpha_k) \Gamma(\sum_{k=1}^{K} n_{m,-(m,n)}^k + \alpha_k)}{\Gamma(n_{m,-(m,n)}^k + \alpha_k) \Gamma(\sum_{k=1}^{K} n_m^k + \alpha_k)}$$

$$\propto \frac{n_{k,-(m,n)}^t + \beta_t}{\sum_{t=1}^{V} (n_{k,-(m,n)}^t + \beta_t)} \cdot \frac{n_{m,-(m,n)}^t + \alpha_k}{\left[\sum_{k=1}^{K} (n_m^t + \alpha_k)\right] - 1}$$

$$(5.24)$$

where $z_{(m,n)}$ is the topic of the n-th word in the m-th document, $z_{-(m,n)}$ is the topics of the other words, $n_{m,-(m,n)}^t$ is the number of the words t in document d_m with $w_{m,n}$ removed, and $n_{k,-(m,n)}^t$ is the number of the words t assigned to the topic k with $w_{m,n}$ removed. Equation (5.24) expresses the probability that the word $w_{m,n}$ is assigned to the topic k. Since this probability is not normalized, we should ensure that all K topics are calculated. Therefore, the time complexity of sampling a new topic for each word is $O(K)$. For each iteration of the Gibbs sampling, the topics of all words in the corpus are updated. Assuming that only a single Markov chain is used, after passing through the burn-in stage, the parameters Θ and Φ can be estimated by using the following equations. The proportion of the topic k in d_m is given as follows:

$$\hat{\theta}_{m,k} = \frac{n_{m,-(m,n)}^k + \alpha_k}{\sum_{k=1}^{K} (n_{m,-(m,n)}^k + \alpha_k)}$$

$$(5.25)$$

The proportion of the word t in the topic k is

$$\hat{\phi}_{k,t} = \frac{n^t_{k,-(m,n)} + \beta_t}{\sum_{t=1}^{V}(n^t_{k,-(m,n)} + \beta_t)} \tag{5.26}$$

Algorithm 3 presents the whole process of applying Gibbs sampling to LDA.

Algorithm 3: Applying Gibbs sampling to LDA

---

1 set the counters n^k_m, n^t_k, n_k to 0
2 **for** *each document d_m, $m \in [1, M]$* **do**
3   **for** *each position in the document d_m, $n \in [1, N_m]$* **do**
4   randomly sample a topic $z_{m,n}$;
5   update the document-topic counter: $n^k_m + 1$;
6   update the topic-word counter: $n^t_k + 1$;
7   update the topic counter:$n_k + 1$;
8   **end**
9 **end**
10 **for** *each iteration* **do**
11   **for** *each document d_m, $m \in [1, M]$* **do**
12   **for** *each position in the document m, $n \in [1, N_m]$* **do**
13   identify the old topic k and the corresponding word t;
14   update the corresponding counter: $n^k_m - 1; m^t_k - 1; n_k - 1$;
15   sample a new topic with reference to Equation (5.24):
                    $k' \sim p(z_{(m,n)} = k | z_{m,-(m,n)}, \boldsymbol{w}; \alpha, \beta)$;
16   update the corresponding counter with the new topic: $n^{k'}_m + 1; n^t_{k'} + 1; n_{k'} + 1$;
17   **end**
18   **end**
19 **end**
20 estimate Φ and Θ according to equations (5.26) and (5.25);

5.5   Metropolis–Hastings Sampling with Alias Method

As described above, the complexity of sampling a topic for a word by Gibbs sampling is linear in the number of topics. Researchers have developed various strategies to improve sampling efficiency. For example, FastLDA [3] reduces time complexity using Hoeffding's inequality. SparseLDA [5] improves the data structure and reduces the complexity of sampling to $O(k_d + k_w)$, where k_d is the number of topics contained in a document d and k_w is the number of topics assigned to w. AliasLDA [1] introduces the alias method to reduce complexity further. LightLDA [6] uses the Metropolis–Hastings sampling algorithm to reduce the complexity to $O(1)$ per word. In this section, we focus on discussing the methods in LightLDA.

5.5.1 Metropolis–Hastings Sampling

Algorithm 4 presents the Metropolis–Hastings algorithm. For a target distribution $p(x)$, the Metropolis–Hastings sampling designs a proposal distribution $q(x'|x)$ that is easier to sample than $p(x)$. The proposal distribution is used to generate a new sample x' based on the current state x. Then the Markov chain decides whether to transfer to x' or remain x, according to the acceptance rate, which is denoted as $A(x, x')$ and calculated as follows:

$$A(x, x') = \min\left\{1, \frac{p(x')}{p(x)}\frac{q(x|x')}{q(x'|x)}\right\} \tag{5.27}$$

It is easy to see that Gibbs sampling is the Metropolis–Hastings sampling with the acceptance rate always being 1.

Algorithm 4: Metropolis–Hastings sampling

---
1 select an initial state x_0
2 set $t = 0$
3 **for** $t = 0, 1, 2, \ldots$ **do**
4   generate a candidate state x' from the proposal distribution $q(x'|x_t)$
5   calculate the acceptance ratio $A(x, x') = \min\{1, \frac{p(x')}{p(x)}\frac{q(x|x')}{q(x'|x)}\}$
6   randomly generate a number u from the uniform distribution $[0, 1]$
7   **if** $u \leq A(x, x')$ **then**
8 | accept the state and set $x_{t+1} = x'$;
9   **else**
10 | reject the state and set $x_{t+1} = x_t$;
11   **end**
12   let $t = t + 1$
13 **end**

A good $q(x'|x)$ increases the sampling efficiency in two ways. First, sampling from $q(x'|x)$ is more convenient than sampling from $p(x)$. Second, an effective $q(x'|x)$ makes the Markov chain converge quickly. However, designing a good proposal distribution is a challenging task. If $q(x'|x)$ and $p(x)$ are mathematically close, sampling from $q(x'|x)$ becomes as difficult as sampling from $p(x)$. If $q(x'|x)$ and $p(x)$ are quite different, sampling from $q(x'|x)$ may be easier, but the acceptance ratio will be low, and the Markov chain will converge slowly.

5.5.2 Alias Method

A uniform distribution can be sampled with the time complexity of $O(1)$. However, sampling from a multinomial distribution with K dimensions typically has a time complexity of $O(K)$. The alias method is an algorithm that effectively improves the sampling efficiency of a multinomial distribution [4].

As shown in Fig. 5.4a, four events (i.e., A, B, C, D) need to be sampled according to their probabilities ($\frac{1}{2}$, $\frac{1}{3}$, $\frac{1}{12}$, $\frac{1}{12}$), and the height of each column visualizes the probability of the corresponding event. Figure 5.4b shows the simplest sampling mechanism. First, the four columns are normalized according to the maximum probability of the four events. Next, one of the four columns is selected at random. Then, a random number ranging from 0 to 1 is generated to determine whether resampling is needed. For example, the event A is sampled if the first column is selected. If the second column is selected and the random number is less than $\frac{2}{3}$, the event B is sampled; otherwise, the sampling fails, and resampling is needed. The complexity of the above algorithm is $O(K)$, where K is the total number of events.

(a) The initial distribution to be sampled from

(b) The simplest sampling mechanism

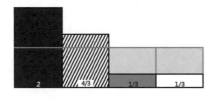

(c) The alias method: step one

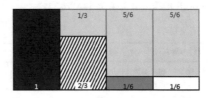

(d) The alias method: step two

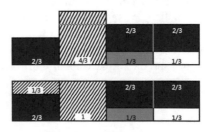

(e) The alias table

Fig. 5.4 Sampling from a multinomial distribution

The alias method transforms the multinomial distribution into a uniform distribution to achieve an amortized time complexity of $O(1)$. In the alias method, there are at most two events in each column. In the first step, the probability of each event is multiplied by K. The result is shown in Fig. 5.4c. Then, in the second step, as shown in Fig. 5.4d and e, the fractions of the first two columns are removed and added to the last two columns, and the first column is further filled with a fraction of the second column. In the alias table, the bottom rectangle corresponds to the initial event probability, while the top rectangle represents the probabilities of the other events.

The implementation of the alias method is based upon two arrays: the prob array and the alias array. The alias table in Fig. 5.4e can be represented by these two arrays. The prob array stores the proportions of the events. In the above example, Prob $= [\frac{2}{3}, 1, \frac{1}{3}, \frac{1}{3}]$. The alias array stores the newly filled events in each column, Alias $= [2, \text{NULL}, 1, 1]$, where NULL indicates that the column contains no additional event. During the sampling process, an integer i is first generated in the range $1 \sim K$ to select a column. Following this, a number is generated in the range $0 \sim 1$. If this number is less than Prob[i], then the event i is sampled. Otherwise, the event Alias[i] is sampled.

The construction of alias table has a time complexity of $O(K)$, while sampling through an alias table has a time complexity of $O(1)$. If the alias table does not require frequent updating, the amortized time complexity is $O(1)$. As such, the alias method is a typical space-for-time algorithm. Alias table reduces the time complexity of sampling from a multinomial distribution by reusing the information stored in table for many times. However, since such information is outdated, the quality of topic models will be significantly impacted if the alias table is not timely updated. In practice, a counter can be maintained for each alias table. When the information of an alias table is used, the corresponding counter is increased by one. When this counter reaches a certain threshold, the table is considered to be obsolete and is updated with the latest information.

5.5.3 Application of Metropolis–Hastings Sampling in LDA

This section discusses how to collectively apply the alias method and the Metropolis–Hastings sampling for LDA. The conditional probability of sampling the i-th word of a document d to the topic z_{di} is

$$
p(z_{di} = k | z_{-(m,n)}, \boldsymbol{w}; \alpha, \beta) \propto \frac{(n_{kd}^{-di} + \alpha_k)(n_{kw}^{-di} + \beta_w)}{n_k^{-di} + \bar{\beta}} \tag{5.28}
$$

The above equation can be approximated as follows:

$$q(z_{di} = k | \boldsymbol{z}_{-(m,n)}, \boldsymbol{w}; \alpha, \beta) \propto \underbrace{(n_{kd} + \alpha_k)}_{\text{function of documents}} \underbrace{\left(\frac{n_{kw} + \beta_w}{n_k + \bar{\beta}} \right)}_{\text{function of words}} \tag{5.29}$$

q is an approximation of the original p. Since q does not require frequent updates, it is more suitable for constructing alias tables. The right side of Eq. (5.29) contains two terms, and each of them represents an effective proposal distribution. Applying alias method to proposal distribution can reduce the sampling complexity to $O(1)$ per word. Although either proposal distribution can be used individually in the Metropolis–Hastings algorithm, LightLDA uses the two proposal distributions alternately to update z_{di}. The first proposal distribution is the word proposal p_w, which is defined as

$$p_w(k) \propto \frac{(n_{kw} + \beta_w)}{n_k + \bar{\beta}} \tag{5.30}$$

The acceptance probability of the transition from a state s to a state t is

$$\min \left\{ 1, \frac{p(t)p_w(s)}{p(s)p_w(t)} \right\} \tag{5.31}$$

Let $\pi_w = \frac{p(t)p_w(s)}{p(s)p_w(t)}$. We then obtain

$$\pi_w = \frac{(n_{td}^{-di} + \alpha_t)(n_{tw}^{-di} + \beta_w)(n_s^{-di} + \bar{\beta})(n_{sw} + \beta_w)(n_t + \bar{\beta})}{(n_{sd}^{-di} + \alpha_s)(n_{sw}^{-di} + \beta_w)(n_t^{-di} + \bar{\beta})(n_{tw} + \beta_w)(n_s + \bar{\beta})} \tag{5.32}$$

Once the sample $t \sim p_w(t)$ is obtained, the acceptance ratio can be calculated in $O(1)$. The second proposal distribution is the document proposal p_d, which is defined as

$$p_d(k) \propto (n_{kd} + \alpha_k) \tag{5.33}$$

The acceptance probability of the transition from a state s to a state t is

$$\min \left\{ 1, \frac{p(t)p_d(s)}{p(s)p_d(t)} \right\} \tag{5.34}$$

Let $\pi_d = \frac{p(t)p_d(s)}{p(s)p_d(t)}$. We then obtain

$$\pi_d = \frac{(n_{td}^{-di} + \alpha_t)(n_{tw}^{-di} + \beta_w)(n_s^{-di} + \bar{\beta})(n_{sd} + \alpha_s)}{(n_{sd}^{-di} + \alpha_s)(n_{sw}^{-di} + \beta_w)(n_t^{-di} + \bar{\beta})(n_{td} + \alpha_t)} \tag{5.35}$$

Similar to the word proposal distribution, the acceptance ratio can be calculated in $O(1)$. Note that in the implementation of LightLDA, the document proposal distribution does not need to build alias table explicitly. Instead, sampling according to the document proposal distribution can be easily achieved by randomly sampling a word in the current document and returning its corresponding topic. Algorithm 5 presents the application of the Metropolis–Hastings sampling to LDA.

Algorithm 5: Applying Metropolis–Hastings sampling and alias method to LDA

---

1 set the counters n_m^k, n_k^t, n_k to 0
2 **for** *each document d_m, $m \in [1, M]$* **do**
3   **for** *each position in the document d_m, $n \in [1, N_m]$* **do**
4   randomly sample a topic $z_{m,n}$
5   update the document-topic counter: $n_m^k + 1$
6   update the topic-word counter: $n_k^t + 1$
7   update the topic counter: $n_k + 1$
8   **end**
9 **end**
10 **for** *each iteration* **do**
11   **for** *each document d_m, $m \in [1, M]$* **do**
12   **for** *each position in the document d_m, $n \in [1, N_m]$* **do**
13   identify the old topic k and the corresponding word t at this position
14   update the corresponding counter: $n_m^k - 1; m_k^t - 1; n_k - 1$
15   **for** *each round of inner iteration* **do**
16   according to a document-topic-proposal distribution, sample a new topic: $k' \sim p(z_{(m,n)} = k | z_{m,-(m,n)}, \boldsymbol{w}; \alpha, \beta)$
17   calculate the acceptance ratio of the topic according to Equation (5.35) and decide whether to accept it or not
18   according to a topic-word-proposal distribution, sample a new topic: $k' \sim p(z_{(m,n)} = k | z_{m,-(m,n)}, \boldsymbol{w}; \alpha, \beta)$
19   calculate the acceptance ratio of the topic according to Equation (5.32) and decide whether to accept it or not
20   **end**
21   update the corresponding counter with the new topic: $n_m^{k'} + 1; n_{k'}^t + 1; n_{k'} + 1$
22   **end**
23   **end**
24 **end**
25 estimate Φ and Θ according to Equation (5.26) and (5.25);

References

1. Li AQ, Ahmed A, Ravi S, Smola AJ (2014) Reducing the sampling complexity of topic models. In: Proceedings of the 20th ACM SIGKDD International Conference on Knowledge Discovery and Data Mining, pp 891–900
2. Nguyen VA, Boyd-Graber J, Resnik P (2014) Sometimes average is best: the importance of averaging for prediction using MCMC inference in topic modeling. In: Proceedings of the 2014 Conference on Empirical Methods in Natural Language Processing (EMNLP), pp 1752–1757
3. Porteous I, Newman D, Ihler A, Asuncion A, Smyth P, Welling M (2008) Fast collapsed Gibbs sampling for latent Dirichlet allocation. In: Proceedings of the 14th ACM SIGKDD International Conference on Knowledge Discovery and Data Mining. ACM, New York, pp 569–577
4. Walker AJ (1974) New fast method for generating discrete random numbers with arbitrary frequency distributions. Electron Lett 10(8):127–128
5. Yao L, Mimno D, Mccallum A (2009) Efficient methods for topic model inference on streaming document collections. In: ACM SIGKDD International Conference on Knowledge Discovery and Data Mining, Paris, June 28–July, pp 937–946
6. Yuan J, Gao F, Ho Q, Dai W, Wei J, Zheng X, Xing EP, Liu TY, Ma WY (2015) LightLDA: big topic models on modest computer clusters. In: Proceedings of the 24th International Conference on World Wide Web, International World Wide Web Conferences Steering Committee, pp 1351–1361

Chapter 6
Variational Inference

Variational inference has become an important research topic in machine learning [1]. It transforms a posterior reasoning problem into an optimization problem and derives a posterior distribution by solving the optimization problem. The fundamental principles of variational inference are introduced in Sect. 6.1. Then, in Sects. 6.2 and 6.3, we discuss evidence lower bound and the mean field variational inference. In Sect. 6.4, we use LDA as an example to show how variational inference is applied to topic models.

6.1 Mathematical Foundation

Assuming that the observed data are x, the variational inference is to estimate the posterior probability of the latent variable z:$p(z|x)$. An approximate probability distribution $q(z; \xi)$ is firstly constructed on z, where ξ is the parameter of this distribution q. $q(z; \xi)$ is further optimized to approximate $q(z|x)$ by optimizing its parameter ξ. The above process involves two key issues: one is how to measure the difference between $q(z; \xi)$ and $p(z|x)$, and the other is how to design a simple form distribution $q(z; \xi)$. To solve the first problem, the KL divergence is used to measure the difference between $q(z; \xi)$ and $p(z|x)$, denoted as $D_{KL}(p||q)$. As described in Sect. 1.2.8, the KL divergence $D_{KL}(p||q)$ is defined as follows:

$$D_{KL}(p||q) = \sum_{x} p(x) \log \frac{p(x)}{q(x)} \tag{6.1}$$

© The Author(s), under exclusive license to Springer Nature Singapore Pte Ltd. 2023
D. Jiang et al., *Probabilistic Topic Models*,
https://doi.org/10.1007/978-981-99-2431-8_6

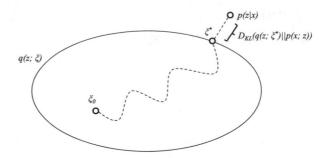

Fig. 6.1 Variational inference

For the second issue, we can theoretically choose any simple distribution $q(z; \xi)$ to approach $q(z|x)$ by optimizing ξ. Then, we transform the problem of estimating $q(z|x)$ into the following optimization problem and find the optimal ξ^* that meets

$$\xi^* = \arg\min_{\xi} D_{KL}\Big[q(z; \xi)||p(z|x)\Big]. \tag{6.2}$$

Figure 6.1 illustrates this process of variational inference. The ellipse represents the possible values of $q(z; \xi)$ with respect to its parameter ξ, after we choose a specific variational distribution q. The dotted line in the figure indicates that some optimization algorithm is used to optimize the parameter ξ in $q(z; \xi)$ iteratively. Finally, when the difference between $q(z|x)$ and $q(z; \xi)$ reaches its minimum, i.e., the minimal $D_{KL}(q(z; \xi)||q(z|x))$ value in the figure, we obtain the optimal value ξ^* of the parameter ξ. $q(z; \xi^*)$ obtained via variational inference is typically called the variational posterior, which can be used as an approximation of the true posterior distribution $q(z|x)$.

6.2   Evidence Lower Bound

The above section introduces the basic concepts of variational inference, but Eq. (6.2) cannot be applied directly because it depends on the true posterior

Fig. 6.2 Minimizing $D_{KL}(q||p)$ by maximizing $\mathcal{L}(\xi)$

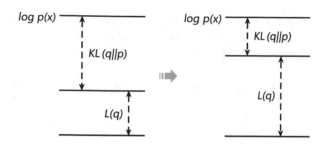

distribution $q(z|x)$. To solve this problem, we conduct the following transformation:

$$
\begin{aligned}
\log p(x) &= \sum_z q(z; \xi) \log p(x) \\
&= \sum_z q(z; \xi) \log \frac{p(x, z)}{p(z|x)} \\
&= \sum_z q(z; \xi) \log \frac{p(x, z)q(z; \xi)}{q(z; \xi)p(z|x)} \\
&= \sum_z q(z; \xi) \log \frac{q(z; \xi)}{p(z|x)} + \sum_z q(z; \xi) \log p(x, z) \\
&\quad - \sum_z q(z; \xi) \log q(z; \xi) \\
&= D_{KL}\Big[q(z; \xi)||p(z|x)\Big] + \mathbb{E}_{q(z;\xi)}\Big[\log p(x, z) - \log q(z; \xi)\Big]
\end{aligned}
\tag{6.3}
$$

where $\log p(x)$ on the left of Eq. (6.3) is called model evidence. The first term on the right is the KL divergence between distributions q and p, and the second term $\mathbb{E}_{q(z;\xi)}\Big[\log p(x, z) - \log q(z; \xi)\Big]$ is called Evidence Lower BOund (ELBO), typically denoted as $\mathcal{L}(\xi)$. As shown in Fig. 6.2, $\log p(x)$ is a constant with respect to the parameter ξ, and we can minimize $D_{KL}(q||p)$ by maximizing $\mathcal{L}(\xi)$.

Through the above transformation, the original optimization problem in Eq. (6.2) is transformed into a new one:

$$
\begin{aligned}
\xi^* &= \arg\max_\xi \mathcal{L}(\xi) \\
&= \arg\max_\xi \mathbb{E}_{q(z;\xi)}\Big[\log p(x, z) - \log q(z; \xi)\Big]
\end{aligned}
\tag{6.4}
$$

where $\mathcal{L}(\xi)$ is the lower bound of the model evidence $\log p(x)$, and optimizing $\mathcal{L}(\xi)$ is equivalent to maximizing the probability of the observed data. According to the definition of $\mathcal{L}(\xi)$, we can decompose it as

$$\mathcal{L}(\xi) = \mathbb{E}_{q(z;\xi)}\Big[\log p(x, z)\Big] - \mathbb{E}_{q(z;\xi)}\Big[\log q(z; \xi)\Big] \tag{6.5a}$$

$$= \mathbb{E}_{q(z;\xi)}\Big[\log p(z)\Big] + \mathbb{E}_{q(z;\xi)}\Big[\log p(x|z)\Big] - \mathbb{E}_{q(z;\xi)}\Big[\log q(z; \xi)\Big] \tag{6.5b}$$

$$= \mathbb{E}_{q(z;\xi)}\Big[\log p(x|z)\Big] - D_{KL}\Big[q(z; \xi)\|p(z)\Big] \tag{6.5c}$$

The first term of Eq. (6.5c) is the expected complete data log-likelihood, and it chooses a variational distribution q explaining observations as much as possible. The second term is the negative KL distance between the variational distribution q and the prior distribution p. The closer these two distributions, the smaller the KL distance and the penalty for $\mathcal{L}(\xi)$ is. Therefore, this equation makes the variational distribution to approximate the actual distribution [3].

6.3   Mean Field Variational Inference

Since the variational inference is a process of maximizing $\mathcal{L}(\xi)$ on a constrained distribution family, the selection of $q(z; \xi)$ has a significant impact on the final result. Both the complexity of calculating $\mathcal{L}(\xi)$ and the approximation of $q(z; \xi)$ and $p(z|x)$ should be considered when selecting $q(z; \xi)$. We can make different assumptions on the relationship between variables in $q(z; \xi)$ to design different $q(z; \xi)$ functions, resulting in different optimizing methods such as the mean field variational method and the structured mean field variational method. We take the mean field variational method as an example in this section. Readers interested in other methods may refer to related studies [4].

In the mean field variational method, we assume that the latent variables in $q(z; \xi)$ are mutually independent, and each of them is represented by a factor. The variational distribution $q(z; \xi)$ can be decomposed into the product of these variational factors:

$$q(z; \xi) = \prod_k q(z_k). \tag{6.6}$$

Using the chain rule, we have

$$p(z_{1:m}, x_{1:n}) = p(x_{1:n}) \prod_{k=1}^{m} p(z_k|z_{1:k-1}, x_{1:n}). \tag{6.7}$$

The expected value of the variational distribution is

$$\mathbb{E}_{q(z;\xi)}\Big[\log q(z_{1:m})\Big] = \sum_{k=1}^{m} \mathbb{E}_{q(z_k)}\Big[\log q(z_k)\Big]. \tag{6.8}$$

Then the ELBO becomes

$$\mathcal{L}(\xi) = \log\ p(x_{1:n}) + \sum_{k=1}^{m} \big(\mathbb{E}_{q(z;\xi)}[\log p(z_k|z_{1:k-1}, x_{1:n})] - \mathbb{E}_{q(z_k)} \log q(z_k)\big). \tag{6.9}$$

The above equation can be regarded as the function of $q(z_k)$. Find the partial derivative of $q(z_k)$ and make the derivative equal to 0:

$$\frac{\mathrm{d}\mathcal{L}}{\mathrm{d}q(z_k)} = \mathbb{E}_{q(z_{-k})}[\log p(z_k|z_{-k}, x)] - \log q(z_k) - 1 = 0. \tag{6.10}$$

Finally, we can get

$$q^*(z_k) \propto \exp\{\mathbb{E}_{q(z_{-k})}[\log p(z_k|z_{-k}, x)]\}. \tag{6.11}$$

The value of z_k can be updated iteratively while fixing the rest z_{-k}. We solve each $q^*(z_k)$ and finally obtain the optimal $q(z; \xi)$. The mean field variational method factorizes complex multivariate integrals into multiple univariate integrals. When the selected $q(z; \xi)$ is the distributions from the exponential family, it can be solved by closed-form solutions.

6.4 Applying Mean Field Variational Inference to LDA

This section takes LDA as an example to introduce how to tackle topic models via variational inference [2, 5], which usually has the following four steps: joint distribution, variational factorization, ELBO, and variational optimization with partial derivatives.

6.4.1 Joint Distribution

Based on the definition of LDA in Chap. 2, we have the joint probability of a word w in the document m, a topic z, the topic distribution θ_m of a document m, and the topic-word distribution Φ as

$$
p(w, z, \theta_m, \Phi | \alpha, \beta) = \underbrace{p(\Phi | \beta)}_{\text{Topic Plate}} \overbrace{p(\theta_m | \alpha) \underbrace{\prod_{n=1}^{N_m} p(w_{m,n} | \phi_{z_{m,n}}) p(z_{m,n} | \theta_m)}_{\text{Word Plate}}}^{\text{Document Plate}} \quad (6.12)
$$

where $\alpha = (\alpha_1, \ldots, \alpha_K)$, $\beta = (\beta_1, \ldots, \beta_V)$. The following discussion is about how to solve the topic distribution θ_m of the document m. We temporarily omit the subscript m and focus on the posterior probability of the latent variables z, θ, and ϕ:

$$
p(z, \theta, \phi | w, \alpha, \beta) = \frac{p(\theta, \phi, z, w | \alpha, \beta)}{p(w | \alpha, \beta)}. \quad (6.13)
$$

However, $p(w | \alpha, \beta)$ involves the following integration:

$$
p(w | \alpha, \beta) = \int \int p(\theta_m | \alpha) p(\Phi | \beta) \prod_{n=1}^{N_m} \sum_{z_{m,n}} p(z_{m,n} | \theta_m) p(w_{m,n} | \phi_{z_{m,n}}) \mathrm{d}\Phi \mathrm{d}\theta_m
$$

$$
= \int \int p(\theta_m | \alpha) p(\Phi | \beta) \prod_{n=1}^{N_m} p(w_{m,n} | \phi_{z_{m,n}}, \Phi) \mathrm{d}\Phi \mathrm{d}\theta_m. \quad (6.14)
$$

Given the coupling among z, θ, and ϕ, it is impossible to get the result by direct integration.

6.4.2   Variational Factorization

Figure 6.3 shows the original graphical model of LDA and the graphical model after introducing the variational factors. To decouple θ, ϕ, and z, new approximate distributions γ, λ, and π are introduced by assuming that θ, ϕ, and z are generated independently. When the coupling relationship between them no longer exists, the mathematical form of the variational distribution q becomes

$$
q(\phi, z, \theta | \gamma, \pi, \lambda) = \prod_{k=1}^{K} q(\phi_k | \lambda_k) \prod_{m=1}^{M} q(\theta_m, z_m | \pi_m, \gamma_m)
$$

$$
= \prod_{k=1}^{K} q(\phi_k | \lambda_k) \prod_{m=1}^{M} \left(q(\theta_m | \gamma_d) \prod_{n=1}^{N_m} q(z_{mn} | \pi_{mn}) \right). \quad (6.15)
$$

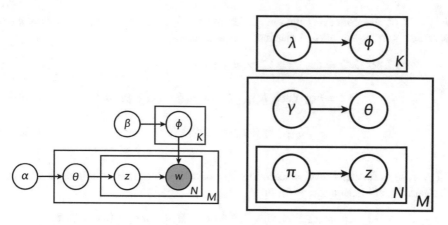

(a) Graphical model of LDA

(b) Graphical model of the variational distribution in LDA

Fig. 6.3 Graphical model of LDA before and after introducing the variational factors

6.4.3 Evidence Lower Bound

After approximating the posterior with variational distributions, the lower bound of the logarithmic likelihood function of the generated document can be derived from the following equation:

$$
\begin{aligned}
\log p(\boldsymbol{w}|\alpha, \beta) &= \log \int \int \sum_z p(\boldsymbol{\phi}, \boldsymbol{\theta}, z, \boldsymbol{w}|\alpha, \beta) \mathrm{d}\theta \mathrm{d}\phi \\
&= \log \int \int \sum_z \frac{p(\boldsymbol{\phi}, \boldsymbol{\theta}, z, \boldsymbol{w}|\alpha, \beta) q(\boldsymbol{\phi}, z, \theta)}{q(\boldsymbol{\phi}, z, \theta)} \mathrm{d}\theta \mathrm{d}\phi \\
&= \log E_{q(\boldsymbol{\phi}, z, \theta)} \left[\frac{p(\boldsymbol{\phi}, \boldsymbol{\theta}, z, \boldsymbol{w}|\alpha, \beta)}{q(\boldsymbol{\phi}, z, \theta)} \right] \\
&\geq E_{q(\boldsymbol{\phi}, z, \theta)} \left[\log \frac{p(\boldsymbol{\phi}, \boldsymbol{\theta}, z, \boldsymbol{w}|\alpha, \beta)}{q(\boldsymbol{\phi}, z, \theta)} \right] \\
&= E_{q(\boldsymbol{\phi}, z, \theta)}[\log p(\boldsymbol{\phi}, \boldsymbol{\theta}, z, \boldsymbol{w}|\alpha, \beta)] - E_{q(\boldsymbol{\phi}, z, \theta)}[\log q(\boldsymbol{\phi}, z, \theta)] \\
&= \mathcal{L}(\gamma, \pi, \lambda) \quad\quad\quad (6.16)
\end{aligned}
$$

In the above equation, we use the variational distribution q to approximate the true posterior. The derivation in the above equation uses the property $f(E(x)) \geq$

$E(f(x))$ of the concave function f. To make q approximate p, we obtain the parameters in the variational distributions as follows:

$$
\begin{aligned}
(\gamma^*, \pi^*, \lambda^*) &= \arg\max_{(\gamma,\pi,\lambda)} \mathcal{L}(\gamma, \pi, \lambda) \\
&= \arg\max_{(\gamma,\pi,\lambda)} E_{q(\boldsymbol{\phi},z,\boldsymbol{\theta})}[\log p(\boldsymbol{\phi}, \boldsymbol{\theta}, z, \boldsymbol{w}|\alpha, \beta)] \\
&\quad - E_{q(\boldsymbol{\phi},z,\boldsymbol{\theta})}[\log q(\boldsymbol{\phi}, z, \boldsymbol{\theta})]
\end{aligned}
\tag{6.17}
$$

$\mathcal{L}(\gamma, \pi, \lambda)$ is further transformed into the expected values of multiple subitems on the variational distribution as follows:

$$
\begin{aligned}
\mathcal{L}(\gamma, \pi, \lambda) &= E_{q(\boldsymbol{\phi},z,\boldsymbol{\theta})}[\log p(\boldsymbol{\phi}, \boldsymbol{\theta}, z, \boldsymbol{w}|\alpha, \beta)] - E_{q(\boldsymbol{\phi},z,\boldsymbol{\theta})}[\log q(\boldsymbol{\phi}, z, \boldsymbol{\theta})] \\
&= E_{q(\boldsymbol{\phi},z,\boldsymbol{\theta})}[\log p(\boldsymbol{\phi}|\beta)p(\boldsymbol{\theta}|\alpha)p(z|\boldsymbol{\theta})p(\boldsymbol{w}|z, \boldsymbol{\phi})] \\
&\quad - E_{q(\boldsymbol{\phi},z,\boldsymbol{\theta})}[\log q(\boldsymbol{\phi}|\lambda)q(z|\pi)q(\boldsymbol{\theta}|\gamma)] \\
&= E_q[\log p(\boldsymbol{\phi}|\beta)] + E_q[\log p(\boldsymbol{\theta}|\alpha)] \\
&\quad + E_q[\log p(z|\boldsymbol{\theta})] + E_q[\log p(\boldsymbol{w}|z, \boldsymbol{\phi})] \\
&\quad - E_q[\log q(\boldsymbol{\phi}|\lambda)] - E_q[\log q(z|\boldsymbol{\phi})] - E_q[\log q(\boldsymbol{\theta}|\gamma)],
\end{aligned}
\tag{6.18}
$$

Note that Dirichlet distribution belongs to the exponential family, and we have:

$$
E[\log \theta_i|\alpha] = \Psi(\alpha_i) - \Psi\left(\sum_{j=1}^k \alpha_j\right)
\tag{6.19}
$$

We further process the seven terms in $\mathcal{L}(\gamma, \pi, \lambda)$.

$E_q[\log p(\boldsymbol{\theta}|\alpha)]$

$$
\begin{aligned}
&= E_q\left[\log\left(\frac{\Gamma(\sum_{k=1}^K \alpha_k)}{\prod_{k=1}^K \Gamma(\alpha_k)}\right)\prod_{k=1}^K \theta_k^{\alpha_k-1}\right] \\
&= E_q\left[\log\Gamma\left(\sum_{k=1}^K \alpha_k\right) - \sum_{k=1}^K \log\Gamma(\alpha_k) + \sum_{k=1}^K (\alpha_k - 1)\log\theta_k\right] \\
&= \log\Gamma\left(\sum_{k=1}^K \alpha_k\right) - \sum_{k=1}^K \log\Gamma(\alpha_k) + \sum_{k=1}^K (\alpha_k - 1)E_q[\log\theta_k] \\
&= \log\Gamma\left(\sum_{k=1}^K \alpha_k\right) - \sum_{k=1}^K \log\Gamma(\alpha_k) + \sum_{k=1}^K (\alpha_k - 1)\left[\Psi(\gamma_k) - \Psi\left(\sum_{k'=1}^K \gamma_{k'}\right)\right].
\end{aligned}
\tag{6.20}
$$

$$E_q[\log p(\boldsymbol{\phi}|\beta)]$$

$$= E_q[\log p(\boldsymbol{\phi}|\beta)]$$

$$= E_q\left[\sum_{k=1}^{K}\log\left(\frac{\Gamma(\sum_{i=1}^{V}\beta_i)}{\prod_{i=1}^{V}\Gamma(\beta_i)}\prod_{i=1}^{V}\phi_{ki}^{\beta_i-1}\right)\right]$$

$$= K\log\Gamma\left(\sum_{i=1}^{V}\beta_i\right) - K\sum_{i=1}^{V}\log\Gamma(\beta_i) + \sum_{k=1}^{K}\sum_{i=1}^{V}(\beta_i-1)E_q[\log\phi_{ki}]$$

$$= K\log\Gamma\left(\sum_{i=1}^{V}\beta_i\right) - K\sum_{i=1}^{V}\log\Gamma(\beta_i)$$

$$+ \sum_{k=1}^{K}\sum_{i=1}^{V}(\beta_i-1)\left[\Psi(\gamma_{ki}) - \Psi\left(\sum_{i'=1}^{V}\gamma_{ki'}\right)\right]. \tag{6.21}$$

$$E_q[\log p(z|\boldsymbol{\theta})]$$

$$= \int\int\sum_z q(\boldsymbol{\phi}, z, \boldsymbol{\theta})\log p(z|\boldsymbol{\theta})\mathrm{d}\theta\mathrm{d}\phi$$

$$= \int\int q(\boldsymbol{\phi}|\lambda)q(\boldsymbol{\theta}|\gamma)\sum_z q(z|\pi)\log p(z|\boldsymbol{\theta})\mathrm{d}\theta\mathrm{d}\phi$$

$$= \int\int q(\boldsymbol{\phi}|\lambda)q(\boldsymbol{\theta}|\gamma)\left(\sum_{z_1=1}^{K}\sum_{z_2-1}^{K}\cdots\sum_{z_n-1}^{K}\prod_{n-1}^{N}q(z_n|\pi_n)\log\prod_{n=1}^{N}p(z_n|\boldsymbol{\theta})\right)\mathrm{d}\theta\mathrm{d}\phi$$

$$= \int\int q(\boldsymbol{\phi}|\lambda)q(\boldsymbol{\theta}|\gamma)\sum_{n=1}^{N}\sum_{k=1}^{K}q(z_n=k|\pi_n)\log p(z_n=k|\boldsymbol{\theta})\mathrm{d}\theta\mathrm{d}\phi$$

$$= \int q(\boldsymbol{\theta}|\gamma)\sum_{n=1}^{N}\sum_{k=1}^{K}\pi_{nk}\log\theta_k\mathrm{d}\theta$$

$$= \sum_{n=1}^{N}\sum_{k=1}^{K}\pi_{nk}E_q[\log\theta_k]$$

$$= \sum_{n=1}^{N}\sum_{k=1}^{K}\pi_{nk}\left(\Psi(\gamma_k) - \Psi(\sum_{k'=1}^{K}\gamma_k')\right). \tag{6.22}$$

$$E_q[\log p(\mathbf{w}|z, \boldsymbol{\phi})]$$

$$= \int \int \sum_z q(\boldsymbol{\phi}, z, \boldsymbol{\theta}) \log p(\mathbf{w}|z, \boldsymbol{\phi}) \mathrm{d}\theta \mathrm{d}\phi$$

$$= \int \int q(\boldsymbol{\phi}|\lambda) q(\boldsymbol{\theta}|\gamma) \left(\sum_{z_1=1}^{K} \sum_{z_2=1}^{K} \cdots \sum_{z_n=1}^{K} \prod_{n=1}^{N} q(z_n|\psi_n) \log \prod_{n=1}^{N} p(w_n|z_n, \phi) \right) \mathrm{d}\theta \mathrm{d}\phi$$

$$= \int \int q(\boldsymbol{\phi}|\lambda) q(\boldsymbol{\theta}|\gamma) \sum_{n=1}^{N} \sum_{k=1}^{K} \sum_{i=1}^{V} q(z_n = k|\psi_n) w_n^i \log p(w_n|z_n = k, \phi) \mathrm{d}\theta \mathrm{d}\phi$$

$$= \sum_{n=1}^{N} \sum_{k=1}^{K} \sum_{i=1}^{V} \pi_{nk} w_n^i E_q[\log \beta_{ki}]$$

$$= \sum_{n=1}^{N} \sum_{k=1}^{K} \sum_{i=1}^{V} \pi_{nk} w_n^i \left(\Psi(\lambda_{ki} - \Psi) \left(\sum_{i'=1}^{V} \lambda_{ki'} \right) \right) \tag{6.23}$$

where w_n^i is 1 only when the nth word is i; otherwise, w_n^i is 0.

$$E_q[\log q(\boldsymbol{\phi}|\lambda)]$$

$$= E_q \left[\sum_{k=1}^{K} \log \left(\frac{\Gamma(\sum_{i=1}^{V} \lambda_{ki})}{\prod_{i=1}^{V} \Gamma(\lambda_{ki})} \prod_{i=1}^{V} \phi_{ki}^{\lambda_{ki}-1} \right) \right]$$

$$= \sum_{k=1}^{K} E_q \left[\log \left(\frac{\Gamma(\sum_{i=1}^{V} \lambda_{ki})}{\prod_{i=1}^{V} \Gamma(\lambda_{ki})} \prod_{i=1}^{V} \phi_{ki}^{\lambda_{ki}-1} \right) \right]$$

$$= \sum_{k=1}^{K} \left(\log \Gamma \left(\sum_{i=1}^{V} \lambda_{ki} \right) - \sum_{i=1}^{V} \log \Gamma(\lambda_{ki}) \right.$$

$$\left. + \sum_{i=1}^{V} (\lambda_{ki} - 1) \left(\Psi(\lambda_{ki}) - \Psi \left(\sum_{i'=1}^{V} \lambda_{ki'} \right) \right) \right). \tag{6.24}$$

$$E_q[\log q(z|\pi)] = E_q\left[\log q(z_1|\pi)q(z_2|\pi)\ldots q(z_n|\pi)\right]$$

$$= \sum_{n=1}^{N} E_q\left[\log q(z_n|\pi_n)\right]$$

$$= \sum_{n=1}^{N} E_q\left[\sum_{k=1}^{K} z_n^k \log \pi_{nk}\right] \qquad (6.25)$$

$$= \sum_{n=1}^{N}\sum_{k=1}^{K} E_q[z_n^k] \log \pi_{nk}$$

$$= \sum_{n=1}^{N}\sum_{k=1}^{K} \pi_{nk} \log \pi_{nk}.$$

$$E_q[\log q(\boldsymbol{\theta}|\gamma)]$$

$$= \log \Gamma\left(\sum_{k=1}^{K}\gamma_k\right) - \sum_{k=1}^{K}\log \Gamma(\gamma_k) + \sum_{k=1}^{K}(\gamma_k - 1)\left(\Psi(\gamma_k) - \Psi\left(\sum_{k'=1}^{K}\gamma_{k'}\right)\right)$$

$$(6.26)$$

6.4.4   *Variational Optimization with Partial Derivatives*

In order to maximize $\mathcal{L}(\gamma, \pi, \lambda)$, the constraint $\sum_{k=1}^{K}\pi_{n,k} = 1$ is introduced to construct the Lagrangian function. Then we calculate the partial derivatives of γ, π, and λ and further obtain the optimized parameters. For simplicity, we only present those related to the derivation parameters for each equation (e.g., $\mathcal{L}_{[\pi]}$ indicates those related to the derivative parameter π):

$$\mathcal{L}_{[\pi]} = \sum_{n=1}^{N}\sum_{k=1}^{K}\pi_{nk}\left(\Psi(\gamma_k) - \Psi\left(\sum_{k'=1}^{K}\gamma_{k'}\right)\right)$$

$$+ \sum_{n=1}^{N}\sum_{k=1}^{K}\sum_{i=1}^{V}\pi_{nk}w_n^i\left(\Psi(\lambda_{ki}) - \Psi\left(\sum_{i'=1}^{V}\lambda_{ki'}\right)\right) \qquad (6.27)$$

$$- \sum_{n=1}^{N}\sum_{k=1}^{K}\pi_{nk}\log \pi_{nk} + \sum_{n=1}^{N}\nu_n\left(\sum_{k=1}^{K}\pi_{nk} - 1\right)$$

Find the partial derivative of the above equation with respect to π:

$$
\frac{\partial \mathcal{L}}{\partial \pi_{nk}} = \sum_{i=1}^{V} w_n^i \left(\Psi\left(\lambda_{ki}\right) - \Psi\left(\sum_{ji'=1}^{V} \lambda_{ki'} \right) \right) - \log \phi_{nk} - 1 + v_n + \Psi(\gamma_k)
$$

$$
- \Psi\left(\sum_{k'=1}^{K} \gamma_{k'} \right) \tag{6.28}
$$

Let the above equation be equal to 0:

$$
\pi_{nk} \propto \exp\left(\sum_{i=1}^{V} w_n^i \left(\Psi(\lambda_{ki}) - \Psi\left(\sum_{i'=1}^{V} \lambda_{ki'} \right) \right) + \Psi(\gamma_k) - \Psi\left(\sum_{k'=1}^{K} \gamma_{k'} \right) \right)
$$

$$
\tag{6.29}
$$

Construct the partial Lagrangian function only containing the parameter γ as follows:

$$
\mathcal{L}_{[\gamma]} = \sum_{n=1}^{N} \sum_{k=1}^{K} \pi_{nk} \left(\Psi(\gamma_k) - \Psi\left(\sum_{j=1}^{K} \gamma_j \right) \right)
$$

$$
+ \sum_{k=1}^{K} (\alpha_k - 1) \left(\Psi(\gamma_k) - \Psi\left(\sum_{j=1}^{K} \gamma_j \right) \right)
$$

$$
- \log \Gamma\left(\sum_{k=1}^{K} \gamma_k \right) + \sum_{k=1}^{K} \log \Gamma(\gamma_k) - \sum_{k=1}^{K} (\gamma_k - 1) \left(\Psi(\gamma_k) - \Psi\left(\sum_{k'=1}^{K} \gamma_{k'} \right) \right)
$$

$$
= \sum_{k=1}^{K} \left(\Psi(\gamma_k) - \Psi\left(\sum_{k'=1}^{K} \gamma_{k'} \right) \right) \left(\sum_{n=1}^{N} \pi_{nk} + \alpha_k - \gamma_k \right) - \log \Gamma\left(\sum_{k=1}^{K} \gamma_k \right)
$$

$$
+ \sum_{k=1}^{K} \log \Gamma(\gamma_k) \tag{6.30}
$$

Finding the partial derivative of the above equation with respect to γ and letting the partial derivative be 0, we get the following equation:

$$
\frac{\partial \mathcal{L}}{\partial \gamma_k} = \left(\Psi'(\gamma_k) - \Psi'\left(\sum_{k'=1}^{K} \gamma_{k'} \right) \right) \left(\sum_{n=1}^{N} \pi_{nk} + \alpha_k - \gamma_k \right), \tag{6.31}
$$

$$
\gamma_k = \alpha_k + \sum_{n=1}^{N} \pi_{nk} \tag{6.32}
$$

Analogously, the Lagrangian function only containing the parameter λ is as follows:

$$
\mathcal{L}_{[\lambda]} = \sum_{k=1}^{K}\sum_{i=1}^{V}(\beta_i - 1)\left(\Psi(\lambda_{ki}) - \Psi\left(\sum_{i'=1}^{V}\lambda_{ki'}\right)\right)
$$
$$
+ \sum_{n=1}^{N}\sum_{k=1}^{K}\sum_{i=1}^{V}\pi_{nk}w_n^i\left(\Psi(\lambda_{ki}) - \Psi\left(\sum_{i'=1}^{V}\lambda_{ki'}\right)\right)
$$
$$
- \sum_{k=1}^{K}\left(\log\Gamma\left(\sum_{i=1}^{V}\lambda_{ki}\right) - \sum_{i=1}^{V}\log\Gamma(\lambda_{ki})\right)
$$
$$
- \sum_{k=1}^{K}\sum_{i=1}^{V}(\lambda_{ki} - 1)\left(\Psi(\lambda_{ki}) - \Psi\left(\sum_{i'=1}^{V}\lambda_{ki'}\right)\right) \tag{6.33}
$$

Let the partial derivative with respect to λ be 0:

$$
\frac{\partial\mathcal{L}}{\partial\lambda_{ki}} = \left(\beta_i + \sum_{n=1}^{N}\pi_{nk}w_n^i - \lambda_{ki}\right)\left(\Psi'(\lambda_{ki}) - \Psi'\left(\sum_{i'=1}^{V}\lambda_{ki'}\right)\right), \tag{6.34}
$$

$$
\lambda_{ki} = \beta_i + \sum_{n=1}^{N}\pi_{nk}w_n^i \tag{6.35}
$$

The above formula is for a specific document, and we omit the document index m for simplicity. When we consider all documents in a corpus D, the topic distributions of the documents are independent. The approach of calculating these parameters becomes

$$
\pi_{mnk} \propto \exp\left(\sum_{i=1}^{V}w_n^i\left(\Psi(\lambda_{ki}) - \Psi\left(\sum_{i'=1}^{V}\lambda_{ki'}\right)\right) + \Psi(\gamma_{mk}) - \Psi\left(\sum_{k'=1}^{K}\gamma_{mk'}\right)\right), \tag{6.36}
$$

$$
\gamma_{mk} = \alpha_k + \sum_{n=1}^{N_m}\pi_{mnk}, \tag{6.37}
$$

$$
\lambda_{ki} = \eta_i + \sum_{m=1}^{M}\sum_{n=1}^{N_m}\pi_{mnk}w_{mn}^i \tag{6.38}
$$

Based upon the above discussion, Algorithm 1 presents the whole process of variational inference for LDA. In recent years, researchers have proposed new methods for applying complex Bayesian models to massive data sets, such as stochastic optimization. Readers can refer to relevant studies [3].

Algorithm 1: Variational inference for LDA

---

1 Input: a dataset D, the total number of topics K, the vocabulary with a size $|V|$.

2 Output: γ^*, λ^* and π^*.

3 Initialize parameters:

4 For all m, n and k, $\pi_{mnk}^{(0)} = \frac{1}{K}$

5 For all m and k, $\gamma_{mnk}^{(0)} = \alpha_k + \frac{N_m}{K}$

6 For all k and i, $\lambda_{ki}^{(0)} = \beta_i + \frac{1}{V}$;

7 while *the convergence condition is not reached* **do**

8   **for** *each document d_m, $m \in [1, M]$* **do**

9   **for** *each word w_n, $n \in [1, N_m]$* **do**

10   **for** *each topic k, $k \in [1, K]$* **do**

11 $\pi_{m,n,k}^{(t+1)} \leftarrow \exp\left(\sum_{i=1}^{V} w_n^i (\Psi(\lambda_{ki}^{(t)}) + \Psi(\gamma_{mk}^{(t)}))\right)$

12   **end**

13   Normalizing $\pi_{m,n,k}^{(t+1)}$;

14   **end**

15 $\gamma_m^{(t+1)} \leftarrow \alpha + \sum_n^N \pi_{m,n}^{(t+1)}$;

16   **end**

17   **for** *each topic k, $k \in [1, K]$* **do**

18   **for** *each word i, $i \in [1, V]$* **do**

19 $\lambda_{k,i}^{(t+1)} \leftarrow \beta_i + \sum_{m=1}^{M} \sum_{n=1}^{N_m} \pi_{m,n,k}^{(t+1)} w_{m,n}^i$

20   **end**

21   **end**

22 $t = t + 1$;

23 end

6.5   Comparison of Variational Inference and MCMC

Compared with MCMC, variational inference has the following properties:

- Variational inference is inherently biased since it optimizes an approximate distribution of the posterior.
- Variational inference is usually faster than MCMC, and MCMC converges slowly to the posterior.
- Variational inference is more friendly to parallelization.

References

1. Bishop CM, Nasrabadi NM (2006) Pattern recognition and machine learning, vol 4. Springer, Berlin
2. Blei DM, Ng AY, Jordan MI (2003) Latent Dirichlet allocation. J Mach Learn Res 3(Jan):993–1022
3. Blei D, Ranganath R, Mohamed S (2016) Variational inference: foundations and modern methods. In: NIPS Tutorial

4. Bouchard-Côté A, Jordan MI (2012) Optimization of structured mean field objectives. Preprint. arXiv:12052658
5. Hoffman MD, Blei DM, Wang C, Paisley J (2013) Stochastic variational inference. J Mach Learn Res 14(1):1303–1347

Chapter 7
Distributed Training

Discovering long-tail topics from massive data usually requires a large number of topics and a large-scale vocabulary. However, using a single machine to train such large-scale topic models encounters bottlenecks in computing efficiency and data storage. Therefore, it is necessary to develop distributed training mechanisms for topic models. In this chapter, we introduce distributed computing architectures in Sect. 7.1, followed by the distributed sampling algorithm in Sect. 7.2, and the distributed variational inference in Sect. 7.3.

7.1 Distributed Computing Architectures

This section covers two common distributed computing architectures: MapReduce and ParameterServer.

7.1.1 MapReduce

MapReduce [1] is a programming paradigm proposed by Google for processing large-scale data sets. The programs that conform to this programming paradigm can be executed in parallel on the MapReduce cluster composed of commodity machines. The execution process of MapReduce is illustrated in Fig. 7.1, whereby the input data are divided into M copies. The Mapper processes the data in parallel on different machines to generate intermediate results in the format of <key,value>. According to the key, the intermediate results are divided into R copies and stored on the disk. The Reducer requests the data on the disk through remote procedure calls and sorts the data according to the key. The intermediate results with the same key are aggregated and allocated to the same machine for further processing.

© The Author(s), under exclusive license to Springer Nature Singapore Pte Ltd. 2023
D. Jiang et al., *Probabilistic Topic Models*,
https://doi.org/10.1007/978-981-99-2431-8_7

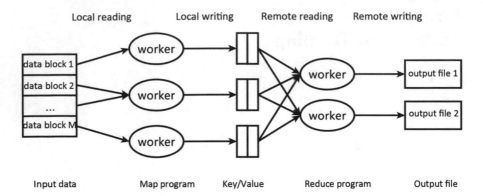

Fig. 7.1 The pipeline of MapReduce

Since MapReduce is responsible for data segmentation, program scheduling, fault tolerance, and communication between machines, software engineers can develop large-scale distributed programs without the expertise of distributed computing.

7.1.2 *ParameterServer*

Many machine learning algorithms need hundreds of iterations to achieve convergence. Since there is considerable variance in data distribution and performance of machines in the cluster, the need for strict synchronization restricts MapReduce's efficiency in distributed machine learning. To alleviate this problem, researchers proposed ParameterServer [2], which supports distributed storage, training, and flexible synchronization mechanisms.

As shown in Fig. 7.2, in ParameterServer, the training data are split into multiple blocks and distributed to multiple workers. In each worker, there is a task scheduler that is responsible for scheduling tasks and monitoring status. The task scheduler allocates the task when a worker either completes the current one or makes an error. The workers do not communicate with each other, and they only communicate with servers. To eliminate the memory bottleneck of each single computing node, the model can be split into multiple blocks and distributed on the servers, which are able to communicate with each other.

In the machine learning task, each worker requests the corresponding parameters from the server, performs local training, records the update of parameters, and transmits the update back to the server for the global update of the model. Since parameter pull, parameter push, and model training can be conducted simultaneously on ParameterServer, there is significant improvement in the efficiency of training large-scale machine learning models. Since the models are distributed in different computing nodes, the synchronization strategy of model parameters is essential for the performance of ParameterServer. There are three modes of

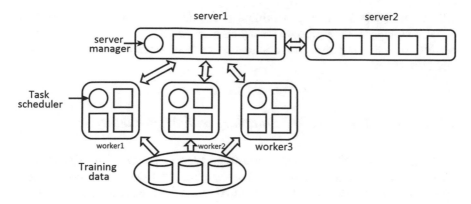

Fig. 7.2 The architecture of ParameterServer

parameter synchronization: bulk synchronous parallel (BSP), asynchronous parallel (ASP), and stale synchronous parallel (SSP):

- In BSP mode, when a worker completes the current task, it waits for the other workers to complete their tasks, and then all the local updates are synchronized to update the global model. BSP ensures that each node synchronizes the parameter updates of the other workers before the next iteration. Since the computing performance of different workers varies and worker failure occurs, the training progress of some workers may lag behind others. This phenomenon limits the overall training efficiency.
- In ASP mode, ParameterServer asynchronously synchronizes the parameters. After the completion of the current task, each worker pushes the update of model parameters to the server, obtains the latest model parameters from the server, and immediately starts the next iteration. Compared with BSP mode, ASP does not require a worker to wait for all other workers to complete the current iteration, and the training efficiency can be significantly improved. However, the model parameters are not consistent across different workers. Faster workers may obtain the old parameters of the other workers, which may lead to compromised training quality and the need for more iterations to achieve convergence.
- SSP is a mode between BSP and ASP. SSP sets a threshold for the iteration delay between workers. This strategy alleviates the straggler problem in BSP and improves training quality. In SSP mode, different distributed algorithms can flexibly utilize different thresholds to balance efficiency and quality.

7.2 Distributed MCMC Sampling

In this section, we utilize LDA as an example to illustrate how to conduct the MCMC sampling algorithms in a distributed approach. The difficulty of parallelizing the MCMC sampling algorithms is that these algorithms are theoretically serial. Researchers have proposed many schemes to relax the serial constraints of MCMC sampling algorithms. We introduce these schemes based on MapReduce and ParameterServer.

7.2.1 Distributed MCMC Sampling with MapReduce

Researchers proposed approximate distributed LDA (AD-LDA) [3], which can be considered as an approximation of Gibbs sampling for LDA at a single machine. Algorithm 1 shows the overall process of AD-LDA. The training data are equally allocated to P computing nodes, and the counters n_w^k and n_k are stored in the global memory. When each iteration starts, each node copies the counters from the global memory to its local memory and denotes them as n_{wp}^k and n_{kp}. Then, each node independently executes Gibbs sampling and saves the intermediate information to the local memory. When each node p completes the current iteration, it updates its local counters n_{wp}^k and n_{kp}. After all the nodes finish the current iteration, the counters on different nodes need to be combined to update the global model: $n_w^k = n_w^k + \sum_{p \in P} (n_{wp}^k - n_w^k), n_k = \sum_{w \in W} n_w^k$. The above process stops when the convergence condition is achieved.

$$p(z_{m,n} = k | z_{-(m,n)}, w; \alpha, \beta) \propto \frac{\left(n_{m,-(m,n)}^k + \alpha_k\right)\left(n_{w,-(m,n)}^k + \beta_w\right)}{n_{k,-(m,n)} + \overline{\beta}} \qquad (7.1)$$

According to the Gibbs sampling formula shown above, when a topic is sampled for the word w in the document m, we need to know the following counts, which depend on the sampling results from the other computing nodes:

1. n_w^k: The count of the word w assigned to the topic k in the whole corpus
2. n_k: The number of words assigned to the topic k in the corpus

When different nodes are processing different words, the sampling results are similar to serial sampling because n_w^k is accurate, and only n_k cannot be timely updated. AD-LDA conducts the global synchronization after each iteration. Experiments show that the delay in updating operation has little influence on the model quality.

   AD-LDA is implemented based on Hadoop. Each node uses the Mapper to conduct Gibbs sampling independently and the Reducer to update global parameters. Since Gibbs sampling requires multiple iterations to converge, the intermediate data are written to HDFS after each iteration. This I/O operation is time-consuming and accounts for nearly 90% of the training time. To further improve the training efficiency, researchers proposed Spark-LDA [4] that stores the intermediate results on the RDD in memory and is $10 \sim 100$ times faster than the counterparts based on Hadoop.

Algorithm 1: AD-LDA

---

1 allocate documents in the training corpus to P computing nodes

2 calculate n_w^k and n_k

3 copy the above counters to each node and record them as n_{wp}^k and n_{kp}

4 **while** *the model does not reach the convergence condition* **do**

5   **for** *each node $p \in (1, 2, \ldots, P)$* **do**

6   (Map phase) execute the single machine Gibbs sampling algorithm, update parameters n_{wp}^k and n_{kp}

7   (Map phase) update the local parameter to the global model:
$$n_w^k = n_w^k + \sum_{p \in P} (n_{wp}^k - n_w^k)$$

8   **end**

9   (Reduce phase) update the global parameter $n_k = \sum_{w \in W} n_w^k$ to check whether the model reaches the convergence condition

10 end

7.2.2 Distributed MCMC Sampling with ParameterServer

A typical ParameterServer solution for training LDA is LightLDA [5]. As shown in Fig. 7.3, multiple servers collectively store the model parameters. The worker requests the parameters needed for the current iteration from the corresponding servers. The communication between the worker and the server depends on ZMQ or MPI. After receiving the worker's request, the corresponding server pushes the parameters to the worker, and the worker trains the model based on these parameters. During the training process, the worker is responsible for aggregating local parameter updates. When the aggregated updates reach a certain amount, they are pushed back to the corresponding server to update the global parameters.

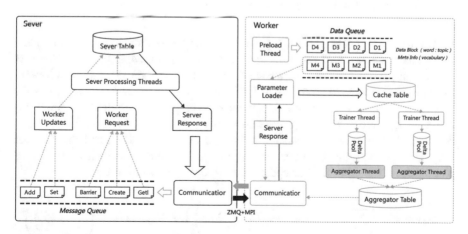

Fig. 7.3 Training LDA with ParameterServer

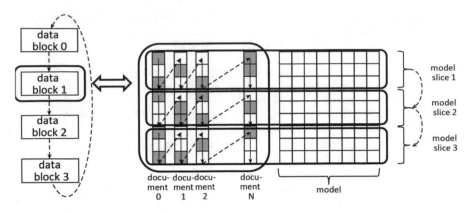

Fig. 7.4 Data blocks and slices of LightLDA

Since LightLDA supports data parallelism and model parallelism, it can train topic models with a large vocabulary and many topics. To optimize memory usage, LightLDA utilizes data blocks and model slices. As shown in Fig. 7.4, the document is divided into several data blocks for each worker. Each data block is divided into different slices according to the corresponding vocabulary. During training, for each slice, the worker requests the corresponding parameters from the servers. As each slice only contains a small subset of the vocabulary, the memory consumption of the local worker is moderate. The parameter updates are recorded on the aggregator table, which is responsible for combining parameter updates and significantly reduces the communication overhead.

7.3   Distributed Variational Inference

In this section, we take Mr. LDA [6] as an example to showcase how to conduct variational inference for LDA through MapReduce. As shown in Fig. 7.5a, there are variational parameters π and γ in each document. π_{mnk} indicates the proportion of the nth word belonging to the topic k in the document m, and $\gamma_{m,k}$ indicates the proportion of the topic k in the document m. The global variational parameter is λ, where $\lambda_{k,v}$ represents the proportion of a word v in the topic k. As shown in Fig. 7.5b, in MR.LDA, the Mapper is used to update π and γ, and the Reducer integrates the information of π and γ and uses them to calculate the global parameter λ. The equations for updating π, γ, and λ are as follows:

$$\pi_{mnk} \propto \exp\left(\sum_{i=1}^{V} w_n^i \left(\Psi(\lambda_{ki}) - \Psi\left(\sum_{i'=1}^{V} \lambda_{ki'}\right)\right) + \Psi(\gamma_{mk}) - \Psi\left(\sum_{k'=1}^{K} \gamma_{mk'}\right)\right)$$

$$(7.2)$$

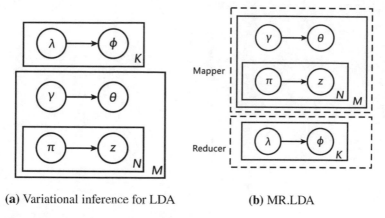

(a) Variational inference for LDA   **(b)** MR.LDA

Fig. 7.5 Variational inference for LDA and MR.LDA

$$\gamma_{mk} = \alpha_k + \sum_{n=1}^{N_m} \pi_{mnk} \qquad (7.3)$$

$$\lambda_{ki} = \eta_i + \sum_{m=1}^{M} \sum_{n-1}^{N_m} \pi_{mnk} w_{mn}^{i} \qquad (7.4)$$

Besides the Mapper and the Reducer, the implementation of MR. LDA includes a Driver, which is responsible for determining the convergence.

References

1. Dean J, Ghemawat S (2008) MapReduce: simplified data processing on large clusters. Commun ACM 51(1):107–113
2. Li M (2014) Scaling distributed machine learning with the parameter server. In: International Conference on Big Data Science and Computing, p 3
3. Newman D, Asuncion A, Smyth P, Welling M (2009) Distributed algorithms for topic models. J Mach Learn Res 10(Aug):1801–1828
4. Qiu Z, Wu B, Wang B, Shi C, Yu L (2014) Collapsed Gibbs sampling for latent Dirichlet allocation on spark. In: Proceedings of the 3rd International Conference on Big Data, Streams and Heterogeneous Source Mining: Algorithms, Systems, Programming Models and Applications, vol 36, pp 17–28
5. Yuan J, Gao F, Ho Q, Dai W, Wei J, Zheng X, Xing EP, Liu TY, Ma WY (2015) LightLDA: Big topic models on modest computer clusters. In: Proceedings of the 24th International Conference on World Wide Web, International World Wide Web Conferences Steering Committee, pp 1351–1361
6. Zhai K, Boyd-Graber J, Asadi N, Alkhouja ML (2012) Mr. LDA: a flexible large scale topic modeling package using variational inference in MapReduce. In: Proceedings of the 21st International Conference on World Wide Web. ACM, New York, pp 879–888

Chapter 8
Parameter Setting

Training high-quality topic models depend on appropriate parameter setting. The parameters in topic models fall into two major categories: hyperparameters and the number of topics. We introduce how to set the hyperparameters in Sect. 8.1, the number of topics in Sect. 8.2, and the evaluation of topic models in Sect. 8.3.

8.1 Hyperparameters

In topic models such as LDA, the hyperparameters refer to the priors of the Dirichlet distributions. Figure 8.1 shows the probability density function of Dirichlet distributions. The values of the hyperparameter α are set to 0.5, 1, and 5. It can be seen that the sparsity of the distribution is determined by the hyperparameters: when α is set to 1, $p(x|\alpha)$ depicts a uniform distribution; when α is set to 5, $p(x|\alpha)$ is a distribution with high density in the central region; when α is set to 0.5, $p(x|\alpha)$ is a sparse distribution with high density in the corner region.

In LDA, the hyperparameters α and β are two smoothing parameters influencing the multinomial distributions Θ and Φ. The sparsity of the multinomial distribution Θ is determined by α, and a small α makes the model represent the document with fewer topics. The sparsity of the multinomial distribution Φ is determined by β, and a small β results in the model assigning fewer words to each topic. Most topic models utilize symmetric Dirichlet prior, assuming that all topics have equal probability in a document and that all words have equal probability in a topic. Empirically, the hyperparameters $\alpha = 50/K$ and $\beta = 0.01$ achieve good performance [2].

© The Author(s), under exclusive license to Springer Nature Singapore Pte Ltd. 2023
D. Jiang et al., *Probabilistic Topic Models*,
https://doi.org/10.1007/978-981-99-2431-8_8

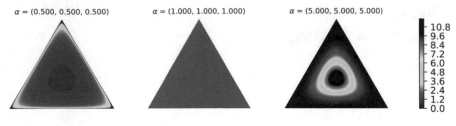

Fig. 8.1 The probability density function of the Dirichlet distribution: $\alpha = 0.5$ in the left figure, $\alpha = 1$ in the central figure, and $\alpha = 5$ in the right figure. The rightmost color bar indicates the density

Besides the value of hyperparameters, researchers have investigated the effect of symmetric and asymmetric hyperparameters [7]. It was found that using an asymmetric Dirichlet prior for the document-topic distribution Θ and a symmetric Dirichlet prior for the topic-word distribution Φ improves the quality of the model. If an asymmetric Dirichlet prior is used for the topic-word distribution Φ, different topics tend to contain similar words, resulting in topical redundancy. If a symmetric Dirichlet prior is used for Φ, the topic-word distribution becomes more balanced without concentrating on a small set of words. As the documents in a specific corpus may be semantically related, different documents can share some topics using an asymmetric Dirichlet prior for the document-topic distribution Θ [7].

Some methods learn the Dirichlet priors based on the training data. However, these methods do not have closed-form solutions and can provide only an approximate solution through iterative optimization. We discuss these methods in the rest of this section.

8.1.1 Hyperparameter Optimization Based on MCMC Sampling

For hyperparameter optimization based on the MCMC sampling results, the maximum likelihood estimation is often used [4]. Given a corpus $D = \{d_m\}_{m=1}^M$, the probability of the Dirichlet prior $\alpha = \{\alpha_k\}_{k=1}^K$ generating the corpus is as follows:

$$
p(D|\alpha) = \prod_{m=1}^M \int p(d_m|\boldsymbol{\theta}_m)p(\boldsymbol{\theta}_m|\alpha)d\boldsymbol{\theta}_m
$$

$$
= \prod_{m=1}^M \left(\frac{\Gamma(\sum_k \alpha_k)}{\Gamma(n_m + \sum_k \alpha_k)} \prod_k \frac{\Gamma(n_m^k + \alpha_k)}{\Gamma(\alpha_k)} \right) \tag{8.1}
$$

where n_m is the total number of words in a document d_m and n_m^k is the total number of words assigned to the topic k in d_m. The log-likelihood of $p(D|\alpha)$ can be calculated as

$$\log p(D|\alpha) = \sum_{m=1}^{M} \left(\log \Gamma \left(\sum_k \alpha_k \right) - \log \Gamma \left(n_m + \sum_k \alpha_k \right) \right.$$

$$\left. + \sum_k \log \Gamma \left(n_m^k + \alpha_k \right) - \sum_k \log \Gamma(\alpha_k) \right) \quad (8.2)$$

The gradient of the log-likelihood can be calculated as

$$g_k = \frac{\partial \log p(D|\alpha)}{\partial \alpha_k}$$

$$= \sum_{m=1}^{M} \left(\Psi \left(\sum_k \alpha_k \right) - \Psi \left(n_m + \sum_k \alpha_k \right) + \Psi \left(n_m^k + \alpha_k \right) - \Psi(\alpha_k) \right)$$

$$(8.3)$$

where $\Psi(x) = d \log(\Gamma(x))/dx$, and it is known as the digamma function. The process of the fixed-point iteration is as follows:

(1) For any $x \in \mathbb{R}^+$ and $a \in \mathbb{Z}^+$, the following inequality holds

$$\log \Gamma(x) - \log \Gamma(x+a) \geq \log \Gamma(\hat{x}) - \log \Gamma(\hat{x}+a) + \left(\Psi(\hat{x}+a) - \Psi(\hat{x}) \right)(\hat{x}-x).$$

(2) For any $x \in \mathbb{R}^+$ and $a \in \mathbb{Z}^+$, the following inequality holds

$$\log \Gamma(x + a) - \log \Gamma(x)$$

$$\geq \log \Gamma(\hat{x} + a) - \log \Gamma(\hat{x}) + \hat{x}\left(\Psi(\hat{x} + a) - \Psi(\hat{x}) \right)(\log x - \log \hat{x}).$$

$$(8.4)$$

Assuming that α^* is the optimal hyperparameter, we have

$$\log P(D|\alpha^*) \geq B(\alpha^*)$$

$$= \sum_{m=1}^{M} \left[\log \Gamma(\alpha) - \log \Gamma(n_m + \alpha) + \left(\Psi(n_m + \alpha) - \Psi(\alpha) \right)(\alpha - \alpha^*) \right.$$

$$+ \sum_k \left[\log \Gamma(n_m^k + \alpha_k) - \log \Gamma(\alpha_k) \right.$$

$$\left. \left. + \alpha_k \left(\Psi(n_m^k + \alpha_k) - \Psi(\alpha_k) \right) \left(\log(\alpha_k^*) - \log(\alpha_k) \right) \right] \right]$$

$$= \sum_{m=1}^{M} \left[\left(\Psi(n_m + \alpha) - \Psi(\alpha) \right)(-\alpha^*) \right.$$

$$\left. + \sum_{k} \alpha_k \left(\Psi(n_m^k + \alpha_k) - \Psi(\alpha_k) \right) \log(\alpha_k^*) \right] + C \tag{8.5}$$

where $B(\alpha^*)$ is the lower bound of $P(D|\alpha^*)$ and C is a constant term. By maximizing the lower bound $B(\alpha^*)$, the maximum likelihood estimate of α_k is as follows:

$$\frac{\partial B(\alpha^*)}{\partial \alpha_k^*} = \sum_{m=1}^{M} \left[\frac{\alpha_k \left(\Psi(n_m^k + \alpha_k) - \Psi(\alpha_k) \right)}{\alpha_k^*} - \left(\Psi(n_m + \alpha) - \Psi(\alpha) \right) \right] = 0 \tag{8.6}$$

We further obtain

$$\alpha_k^* = \frac{\alpha_k \left(\sum_{m=1}^{M} \Psi(n_m^k + \alpha_k) - M\Psi(\alpha_k) \right)}{\sum_{m=1}^{M} \Psi(n_m + \sum_k \alpha_k) - M\Psi(\sum_k \alpha_k)}$$

After several iterations, the optimization process converges and the estimation of α is obtained. The estimation of β can be calculated analogously.

8.1.2 Hyperparameter Optimization Based on Variational Inference

Hyperparameter optimization based on variational inference relies on iterating the following two steps. The E-step optimizes the variational parameters π, λ, and γ; the M-step maximizes the log-likelihood $\mathcal{L}$ by optimizing the hyperparameters α and β:

$$\mathcal{L} = E_{q(\phi,z,\theta)}[\log p(\phi, \theta, z, w|\alpha, \beta)] - E_{q(\phi,z,\theta)}[\log q(\phi, z, \theta)] \tag{8.7}$$

where α and β are optimized using Newton method. The Newton method requires calculating the first and second derivatives of α and β. The first derivative of α is

$$\nabla_{\alpha_i} \mathcal{L} = M \left(\Psi \left(\sum_{j=1}^{k} \alpha_j \right) - \Psi(\alpha_i) \right) + \sum_{d=1}^{M} \left(\Psi(\gamma_{di}) - \Psi \left(\sum_{j=1}^{k} \gamma_{dj} \right) \right) \tag{8.8}$$

The second derivative of α is

$$\nabla_{\alpha_i \alpha_j} \mathcal{L} = \delta(i, j) M \Psi'(\alpha_i) - \Psi' \left(\sum_{j=1}^{k} \alpha_j \right) \tag{8.9}$$

When $i = j$, $\delta(i, j) = 1$; otherwise, $\delta(i, j) = 0$. The update equation of α is

$$\alpha_i \leftarrow \alpha_i + \frac{\nabla_{\alpha_i}\mathcal{L}}{\nabla_{\alpha_i\alpha_j}\mathcal{L}} \tag{8.10}$$

The hyperparameter β can be optimized analogously. The E-step and M-step are iterated until the lower bound of L converges.

8.2 The Number of Topics

The number of topics is crucial for topic models. One of the most common methods to determine the number of topics is to manually select the number of topics in terms of the logarithmic likelihood or perplexity. Another commonly used method is Bayesian nonparametric [6], which automatically determines the number of topics based on the training data. Since predefining the number of topics is more common in real applications, we discuss how to set the number of topics in this section manually.

For topic models with different numbers of topics, the logarithmic likelihood or perplexity can be tested on the validation data set to reflect the quality of the corresponding models. The number of topics with the maximum logarithmic likelihood or the minimum perplexity is considered as the one that best fits the data. For the validation data set D, the logarithmic likelihood with regard to a topic model $\mathcal{M}$ is as follows:

$$
\begin{aligned}
\text{Logarithmic Likelihood}&(D|\mathcal{M}) \\
&= \sum_{m=1}^{M} \log p(d_m|\mathcal{M}) \\
&= \sum_{m=1}^{M}\sum_{n=1}^{n_m} \log(p(w_{m,n}|\mathcal{M})) \\
&= \sum_{m=1}^{M}\sum_{n=1}^{n_m} \log\left(\sum_{k=1}^{K} p(w_{m,n}, z_{m,n}=k|\mathcal{M})\right) \\
&= \sum_{m=1}^{M}\sum_{n=1}^{n_m} \log\left(\sum_{k=1}^{K} p(w_{m,n}|z_{m,n}=k)p(z_{m,n}=k|\mathcal{M})\right) \\
&= \sum_{m=1}^{M}\sum_{n=1}^{n_m} \log\left(\sum_{k=1}^{K} \phi_{k,t}\theta_{m,k}\right)
\end{aligned}
\tag{8.11}
$$

where d_m is the m-th document in the validation set, n_m is the number of words in d_m, and $\theta_{k,m}$ is inferred based on $\mathcal{M}$. The perplexity is defined as the reciprocal of the geometric mean of the likelihood of each word on the validation data set:

$$\text{Perplexity}(D|\mathcal{M}) = \exp\left(-\frac{\sum_{m=1}^{M} \log p(d_m|\mathcal{M})}{\sum_{m=1}^{M} n_m}\right) \qquad (8.12)$$

The relationship between perplexity and logarithmic likelihood is

$$\text{Perplexity}(D|\mathcal{M}) = \exp\left(-\frac{\text{Logarithmic-Likelihood}(D|\mathcal{M})}{\sum_{m=1}^{M} n_m}\right) \qquad (8.13)$$

The above equation shows that the perplexity can be easily calculated based on the logarithmic likelihood.

In practice, topic models are trained with different numbers of topics, and the corresponding logarithmic likelihood or perplexity on the validation data is calculated for each topic model. The one that exhibits maximum logarithmic likelihood or the minimum perplexity is identified as the optimal model.

8.3 Advanced Metrics for Model Evaluation

We have discussed how to evaluate the quality of topic models based upon logarithmic likelihood and perplexity. However, recent studies show that these metrics are not necessarily aligned with the real performance in specific tasks. Hence, some advanced metrics are proposed to evaluate the quality of topic models from new perspectives.

One metric is interpretability. Researchers have designed two tasks (i.e., word intrusion and topic intrusion) to gauge the interpretability of a topic model [1]. The word intrusion task is used to evaluate the semantic consistency of the words in a topic. The top-5 largest weight words are selected from each topic. Then, another word with a small weight in this topic but a large weight in the other topics is chosen as the intrusion word. When the six words are mixed, if the top-5 words in the topic are semantically consistent, the intrusion word would be easy to figure out. However, if the semantic meaning of the topic is unclear, it becomes challenging for human annotators to figure out the intrusion word. The topic intrusion task evaluates whether the topics assigned to a document are aligned with human intuition. The title of a document, a section of the document, and four topics are displayed to human annotators. The annotators need to figure out the intrusion topic. The four topics include the top-3 topics of the document and one low-probability topic as the intrusion topic. If the topic model is of high quality, it becomes straightforward for human annotators to determine the intrusion topic.

The above method relies on human annotation and is inappropriate for large-scale evaluation. Therefore, researchers develop an approach to automatically quantify the semantic consistency of a topic using some external resources [5], such as Wikipedia data and pointwise mutual information (PMI). PMI measures the co-occurrence of words. For any pair of words (w_i, w_j) in the first K words of a topic, the PMI is defined as follows:

$$\text{PMI}(w_i, w_j) = \log \frac{p(w_i, w_j) + \epsilon}{p(w_i)p(w_j)}$$

where ϵ is a smoothing item. $p(w_i, w_j)$, $p(w_i)$, and $p(w_j)$ are estimated based on Wikipedia data. The semantic consistency of a topic is calculated based on the average PMI of the pairwise combinations of the first K words in this topic.

Some researchers utilize the training corpus to evaluate the consistency of topics based on the co-occurrence information at the document level [3]. For any pair of words (w_i, w_j), their scores are

$$\text{score}(w_i, w_j) = \log \frac{D(w_i, w_j) + \epsilon}{D(w_j)},$$

where ϵ is also a smoothing item. $D(w_i, w_j)$ represents the number of documents that contain both w_i and w_j as words, while $D(w_j)$ represents the number of documents that contain the word w_j. The semantic consistency of a topic is calculated based on the average score of the pairwise combinations of words in this topic.

In addition to the above metrics, some other metrics evaluate the quality of the topic model through metrics from specific tasks, such as recall, precision, F1, and normalized discounted cumulative gain (NDCG), etc.

References

1. Chang J, Boyd-Graber JL, Gerrish S, Wang C, Blei DM (2009) Reading tea leaves: how humans interpret topic models. In: Advances in Neural Information Processing Systems, vol 31, pp 1–9
2. Hong L, Davison BD (2010) Empirical study of topic modeling in Twitter. In: Proceedings of the First Workshop on Social Media Analytics. ACM, New York, pp 80–88
3. Mimno D, Wallach HM, Talley E, Leenders M, McCallum A (2011) Optimizing semantic coherence in topic models. In: Proceedings of the Conference on Empirical Methods in Natural Language Processing. Association for Computational Linguistics, Cedarville, pp 262–272
4. Minka T (2000) Estimating a Dirichlet distribution. Technical report, MIT
5. Newman D, Lau JH, Grieser K, Baldwin T (2010) Automatic evaluation of topic coherence. In: Human Language Technologies: The 2010 Annual Conference of the North American Chapter of the Association for Computational Linguistics. Association for Computational Linguistics, Cedarville, pp 100–108
6. Teh YW, Jordan MI, Beal MJ, Blei DM (2006) Hierarchical Dirichlet processes. J Am Stat Assoc 101(476):1566–1581
7. Wallach HM, Mimno DM, McCallum A (2009) Rethinking LDA: why priors matter. In: Advances in Neural Information Processing Systems, pp 1973–1981

Chapter 9
Topic Deduplication and Model Compression

There are two critical issues when using the trained topic models: the redundancy of topics and the high computational costs for topic inference. Therefore, there is still room for further optimizing the trained topic models. In this chapter, we introduce topic deduplication and model compression. These two techniques significantly affect real-life applications [3]. The methods are discussed based on LDA but can be easily transferred to other topic models.

9.1 Topic Deduplication

Topic models are usually plagued with the problem of topic redundancy. This problem is particularly prominent when large-scale topic models are trained on massive data. Figure 9.1 gives an example of topic redundancy. Since the three topics contain words such as "video," "log on," and "download," the similarity between these topics is exceptionally high. If these redundant topics can be aggregated into one, the remaining topics become more independent, and the entire model becomes more interpretable. In this section, we introduce a topic deduplication algorithm, which does not modify the training process and only serves as a downstream processing strategy for topic models.

9.1.1 Precise Topic Deduplication

We use the topic model trained by the sampling algorithms to introduce the topic deduplication, which can be easily transferred to topic models trained by variational inference. Assume the trained topic model is $\{(w_1, c_{w_1}^{z_j}), (w_2, c_{w_2}^{z_j}), \cdots\}$, in which w_i represents the i-th word, z_j represents the j-th topic, and $c_{w_i}^{z_j}$ represents the count

Fig. 9.1 Topic redundancy

Topic 393	Topic 1610	Topic 4855
video: 0.0867	video: 0.2480	video: 0.2697
download:0.0352	log on:0.0926	download:0.1820
log on: 0.0295	download: 0.0716	log on: 0.1690
client-side: 0.0257	check: 0.0629	app: 0.0964
check: 0.0254	user: 0.0431	phone: 0.0964
ad: 0.0234	ad: 0.0369	share: 0.0349
user: 0.0203	content: 0.0368	secretly: 0.0155
record: 0.0180	game: 0.0298	account: 0.0144
game: 0.0168	use: 0.0296	music: 0.0121
VIP: 0.0164	record: 0.0255	more: 0.0093

Fig. 9.2 Topic model in the topic-word format

	z_1	z_2	$\cdots$	z_N
w_1	$p_{w_1}^{z_1}$	$p_{w_1}^{z_2}$	$\cdots$	$p_{w_1}^{z_N}$
w_2	$p_{w_2}^{z_1}$	$p_{w_2}^{z_2}$	$\cdots$	$p_{w_2}^{z_N}$
$\cdots$	$\cdots$	$\cdots$	$\cdots$	$\cdots$
w_M	$p_{w_M}^{z_1}$	$p_{w_M}^{z_2}$	$\cdots$	$p_{w_M}^{z_N}$

of the word w_i assigned to topic z_j. To further obtain the probability distribution of each word, we calculate the number of words under the topic z_j as s_{z_j}. The probability for each word is further calculated as $p_{w_i}^{z_j} = c_{w_i}^{z_j}/s_{z_j}$, and the probability form of the topic-word format is $\mathcal{M} = \{(w_1, \ p_{w_1}^{z_j}), \ (w_2, \ p_{w_2}^{z_j}), \ \cdots\}$, which is shown in Fig. 9.2.

The precise topic deduplication algorithm analyzes the similarity between topics, merges redundant topics, and generates a new topic model. The topic deduplication algorithm primarily includes two stages: topic similarity analysis and topic fusion.

9.1.1.1 Topic Similarity Analysis

We can measure the redundancy of two topics by Jaccard similarity or weighted Jaccard similarity [2] on the first T words.

For two topics z_i and z_j, the Jaccard similarity between them is defined as

$$S(z_i, z_j) = \frac{|W_{z_i} \cap W_{z_j}|}{|W_{z_i} \cup W_{z_j}|} = \frac{|W_{z_i} \cap W_{z_j}|}{|W_{z_i}| + |W_{z_j}| - |W_{z_i} \cap W_{z_j}|} \tag{9.1}$$

where W_{z_i} and W_{z_j} are the lists composed of the first T words of the two topics. The numerator $|W_{z_i} \cap W_{z_j}|$ represents the number of shared words in the first T words of the two topics; the denominator $|W_{z_i} \cup W_{z_j}|$ represents the total number of words covered by the first T words of the two topics. Taking the topics z_{393} and z_{1060} in Fig. 9.1 as an example, we consider the first five words of each topic when analyzing the topic similarity. $W_{393} = \{$video, download, log on, client-side and check$\}$, and $W_{1060} = \{$video, log on, download, check, user$\}$. The number of overlapping words in W_{393} and W_{1060} is $|W_{z_i} \cap W_{z_j}| = 4$, and the number of words covered in W_{393} and W_{1060} is $|W_{z_i} \cup W_{z_j}| = 6$. Hence, the Jaccard similarity of z_{393} and z_{1060} is $S(z_{393}, z_{1060}) = 0.67$.

For any two topics, z_i and z_j, the weighted Jaccard similarity between them is defined as

$$\begin{aligned}
S(z_i, z_j) &= \frac{\sum_1^m \min(p_{w_t}^{z_i}, p_{w_t}^{z_j})}{\sum_1^m \max(p_{w_t}^{z_i}, p_{w_t}^{z_j}) + \sum_{m+1}^T p_{w_t}^{z_i} + \sum_{m+1}^T p_{w_t}^{z_j}} \\
&= \frac{\sum_1^m \min(p_{w_t}^{z_i}, p_{w_t}^{z_j})}{\sum_1^T p_{w_t}^{z_i} + \sum_1^T p_{w_t}^{z_j} - \sum_1^m \min(p_{w_t}^{z_i}, p_{w_t}^{z_j})}
\end{aligned} \tag{9.2}$$

In the weighted Jaccard similarity, $P_{z_i} = (p_{w_1}^{z_i}, p_{w_2}^{z_i}, \cdots, p_{w_m}^{z_i}, p_{w_{m+1}}^{z_i}, \cdots, p_{w_T}^{z_i})$ and $P_{z_j} = (p_{w_1}^{z_j}, p_{w_2}^{z_j}, \cdots, p_{w_m}^{z_j}, p_{w_{m+1}}^{z_j}, \cdots, p_{w_T}^{z_j})$ are the vectors composed of the corresponding probabilities of the first T words of the two topics. m represents the number of words shared by z_i and z_j. The numerator $\sum_1^m \min(p_{w_t}^{z_i}, p_{w_t}^{z_j})$ is the sum of the smaller probability of a shared word in the two topics. The denominator $\sum_1^m \max(p_{w_t}^{z_i}, p_{w_t}^{z_j}) + \sum_{m+1}^T p_{w_t}^{z_i} + \sum_{m+1}^T p_{w_t}^{z_j}$ is the sum of the larger probability of repeated words in both topics and the probability of non-shared words. We take the topics z_{393} and z_{1060} in Fig. 9.1 as an example and consider the first five words of each topic. The topic z_{393} is $\{$(video, 0.087), (download, 0.035), (login, 0.029), (client-side, 0.026), (check, 0.025)$\}$, and the topic z_{1060} is $\{$(video, 0.248), (login, 0.093), (download,0.072), (check, 0.063), (user, 0.043)$\}$. Given the four shared words $\{$video, download, login, checking$\}$, $P_{z_{393}} = (0.087, 0.035, 0.029, 0.025, 0.026)$, and $P_{z_{1060}} = (0.248, 0.072, 0.093, 0.063, 0.043)$. According to the above definition, the numerator is: $0.087 + 0.035 + 0.029 + 0.025 = 0.176$, and the denominator is:

$0.248 + 0.072 + 0.093 + 0.063 + 0.026 + 0.043 = 0.545$. The weighted Jaccard similarity of z_{393} and z_{1060} is calculated as $S(z_{393}, z_{1060}) = 0.323$.

For two topics z_i and z_j, if the similarity is greater than the threshold δ_J set by the user, it is determined that there is a significant redundancy between the two topics, and the redundant topic pairs (z_i, z_j) are recorded in $\mathcal{R}$.

9.1.1.2   Topic Fusion

After all redundant topic pairs (z_i, z_j) are recorded in $\mathcal{R}$, we merge redundant topics through topic fusion. Topic fusion finds redundant topic sets through the union-find algorithm, which is illustrated as follows.

Assume that $\mathcal{R}$ is $\{(1, 2), (2, 3), (4, 5), (6, 7), (1, 7)\}$, where $(1, 2)$ indicates that there is redundancy between topic 1 and topic 2, $(2, 3)$ indicates that there is redundancy between topic 2 and topic 3, and so on. The union-find algorithm first establishes an index vocabulary I_M for topics, in order to efficiently retrieve the index of each topic in $\mathcal{R}$. The I_M built for the above example is: $\{1 : [0, 4], 2 : [0, 1], 3 : 1, 4 : 2, 5 : 2, 6 : [3], 7 : [3, 4]\}$, where $1 : [0, 4]$ indicates that topic 1 appears in $\mathcal{R}[0]$ and $\mathcal{R}[4]$. The sets under the same topic can be regarded as interconnected. For example, $[0, 4]$ in I_M denotes that $\mathcal{R}[0]$ and $\mathcal{R}[4]$ are connected, and $[0, 1]$ denotes that $\mathcal{R}[0]$ and $\mathcal{R}[1]$ are connected. Hence, $\mathcal{R}[0]$, $\mathcal{R}[1]$, and $\mathcal{R}[4]$ are connected. Based on I_M, we get the set $I_C = \{(0, 1, 3, 4), (2)\}$, where in each set the corresponding sets are connected, indicating that the corresponding topics are redundant. In the above example, we get the redundant topic set $I_R = \{(1, 2, 3, 6, 7), (4, 5)\}$, where $(1,2,3,6,7)$ indicates that the topics 1, 2, 3, 6, and 7 are redundant.

Based on I_R, we select the first topic in each set of I_R as the root topic and merge the remaining topics in the same set into the root topic. In the previous example, the redundant topic set obtained by the union-find algorithm is $\{(1, 2, 3, 6, 7), (4, 5)\}$. For the set $(1, 2, 3, 6, 7)$, the topic 1 can be selected as the root topic, the topics 2, 3, 6, and 7 can be merged into topic 1, and the contents of the row of topics 2, 3, 6, and 7 can be deleted from the model. The topic merging here refers to the statistical accumulation of the two topics. After z_j is merged into z_i, $z_i = \{(w_1, p_{w_1}^{z_i} + p_{w_1}^{z_j}), (w_2, p_{w_2}^{z_i} + p_{w_2}^{z_j}), \cdots \}$, in which the probabilities are further normalized. After all redundant topic sets are merged, all empty rows are removed to generate a new topic model in the topic-word format.

To verify the effectiveness of the topic deduplication algorithm, we evaluate the redundancy of the model after topic deduplication under different thresholds. As shown in Fig. 9.3, as the threshold decreases, the redundancy of the generated topic model continues to decrease. To verify the effect of the topic model after deduplication in practical applications, we use the original model without deduplication (5000 topics), the model after deduplication using the Jaccard similarity strategy (4798 topics), and the model after deduplication using the weighted Jaccard similarity strategy (4537 topics) to conduct a comparative experiment on text classification.

Algorithm 1: The union-find algorithm

---

1 begin
2 | initialize the index vocabulary $I_M = \{\}$, the connected set $I_C = \{\}$, the redundant topic set $I_R = \{\}$;
3 | **for** (z_i, z_j) *in* $\mathcal{R}$ **do**
4 | | add the indexes z_i and z_j to I_M;
5 | **end**
6 | merge the indexes in I_M, and obtain the connected set I_C;
7 | replace the indexes in I_C with the corresponding topics;
8 end
9 return I_R;

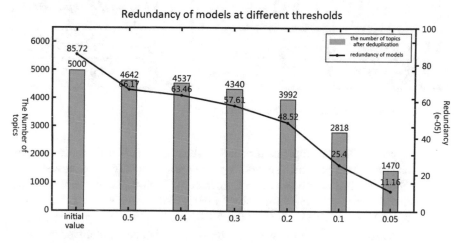

Fig. 9.3 The redundancy of the model under different thresholds δ_J

The initial model, the Jaccard deduplication model, and the weighted Jaccard deduplication model achieve classification accuracies of 89.2%, 89.3%, and 89.7%. We can see that the model after topic deduplication is not only more concise but also improves the performance of downstream applications.

9.1.2 Fast Topic Deduplication

In the above deduplication algorithm, we need to calculate the similarity between any two topics. When the number of topics is huge, the efficiency of this algorithm is extremely low. To improve the efficiency of deduplication on large-scale topic models, we introduce a fast topic deduplication algorithm based on SimHash [1], which is a fast deduplication algorithm for massive texts and represents each document with hash code. It evaluates the similarity of two documents by judging whether the hamming distance of the hash codes of the two documents is less than a

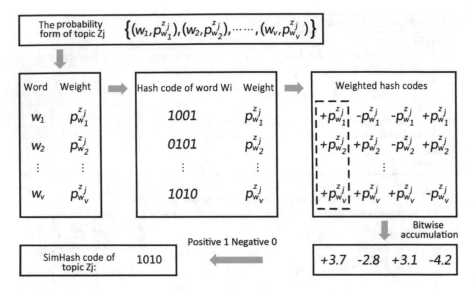

Fig. 9.4 The process of calculating SimHash for a topic

certain threshold. The fast topic deduplication algorithm is divided into two stages: topic clustering based on SimHash and topic deduplication within the cluster.

As shown in Fig. 9.4, we calculate SimHash for each topic. We first calculate the hash code corresponding to the word string. For example, if the hash code corresponding to w_1 is 1001, the weighted hash code of w_1 is $+p_{w_1}^{z_j} - p_{w_1}^{z_j} - p_{w_1}^{z_j} + p_{w_1}^{z_j}$. After obtaining the weighted hash code of each word under the topic, we accumulate them together and further transform them into the SimHash code of the topic. The greater the similarity of two topics, the smaller the calculated hamming distance of the SimHash code. However, the distribution of words in topics has its particularity. Since some words account for an excessive proportion of the distribution, the SimHash codes of many topics have zero hamming distances. Hence, topic fusion that depends only on SimHash may integrate dissimilar topics together. To improve the quality of fast deduplication, we further calculate the Jaccard similarity of the topics in the same cluster. Suppose the similarity of the two topics exceeds the threshold δ_J (set by the user). In that case, it is determined that there is sufficient redundancy between the two topics, and the redundant topic pairs are recorded. These redundant topic pairs will be processed by topic fusion.

As shown in Algorithm 2, the fast deduplication algorithm only estimates the similarity of two topics in each cluster and significantly reduces the computational complexity. Compared with the precise topic duplication, the fast topic duplication can improve efficiency by nearly eight times. We also analyze the redundancy of topic models using different deduplication algorithms and find that fast topic deduplication demonstrates similar effect as the precise topic deduplication.

Algorithm 2: Fast topic deduplication

---

1 **begin**
2   initialize the cluster $C = \{\}$, the redundant topic set $\mathcal{R} = \{\}$;
3   **for** *a topic z_i in M* **do**
4 | calculate the SimHash code of z_i;
5   **end**
6   cluster the topics with the same SimHash code into C;
7   **for** *a set c in C* **do**
8 | calculate the similarity of two topics in c according to Equation (9.1)
         | or (9.2), and add redundant topics to $\mathcal{R}$;
9   **end**
10   **for** *a set s in union-find($\mathcal{R}$)* **do**
11   **for** *a topic z_{sk} ($k > 1$) in s* **do**
12 | add z_{sk} to z_{s1} and delete M in z_{sk};
13   **end**
14   **end**
15    $M^C = M$;
16 **end**
17 **return** M^C;

9.2 Model Compression

This section introduces topic model compression, which makes topic models consume less memory and more efficient at topic inference. We first present an algorithm named topic-dimension compression for the topic model obtained by MCMC sampling or variational inference. Then, we introduce an algorithm called word-dimension compression for efficient topic inference.

9.2.1 Topic-Dimension Compression

Figure 9.5 presents the multinomial distribution of words in a topic. It can be seen that many tail words have very small weights, and these words have little impact on downstream applications. To save storage space and improve the efficiency of topic inference, we compress the model by deleting these low-frequency words.

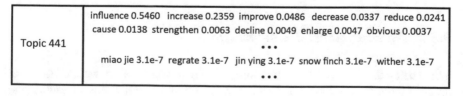

Topic 441	influence 0.5460 increase 0.2359 improve 0.0486 decrease 0.0337 reduce 0.0241
	cause 0.0138 strengthen 0.0063 decline 0.0049 enlarge 0.0047 obvious 0.0037
	• • •
	miao jie 3.1e-7 regrate 3.1e-7 jin ying 3.1e-7 snow finch 3.1e-7 wither 3.1e-7
	• • •

Fig. 9.5 An example of word distribution in a topic

Algorithm 3: Topic-dimension compression

---

1 **begin**
2   **for** *each topic z_j in $\mathcal{M}$* **do**
3   **for** *each word w_i in z_j* **do**
4   **if** $p_{w_i}^{z_j} \leq \delta_T$ **then**
5   delete the items $(w_i, \; p_{w_i}^{z_j})$ from z_j
6   **end**
7   **end**
8   **end**
9 **end**

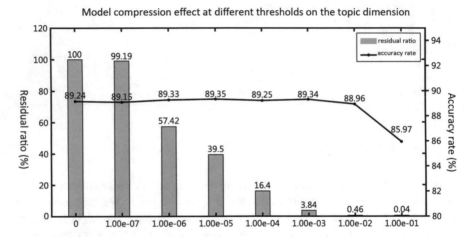

Fig. 9.6 Model compression at different thresholds on the topic dimension

Algorithm 3 illustrates the process of topic-dimension compression. The format of a topic z_j is $\{(w_1, \; p_{w_1}^{z_j}), \; (w_2, \; p_{w_2}^{z_j}), \; \cdots\}$. If the $p_{w_i}^{z_j}$ of w_i is less than a threshold, we delete the corresponding $(w_i, \; p_{w_i}^{z_j})$. To verify the effectiveness of the topic model after compression, we evaluate it on the text classification task. As shown in Fig. 9.6, with the increase of δ_T, the compression strength increases, and the number of remaining items decreases. When the threshold is set to 10^{-5}, 60.05% of the items can be removed, and the topic model after compression can still achieve similar accuracy as the original model.

9.2.2 Word-Dimension Compression

When using the Metropolis–Hastings algorithm based on the alias, we need to create an alias table for each word and calculate the sampling probability of each topic. As the probabilities of sampling certain topics are extremely low, for storage and computational efficiency, we delete these low-frequency topic items to compress the model. The process of word-dimension compression is shown in Algorithm 4. The word-topic proposal for each word is calculated as follows:

$$a_{w_i}^{z_j} = \frac{p_{w_i}^{z_j}}{\sum_{z'} p_{w_i}^{z'}} \tag{9.3}$$

If $a_{w_i}^{z_j}$ is less than a threshold δ_A, we delete the corresponding item $(z_j : p_{w_i}^{z_j})$ and execute normalization again.

Algorithm 4: Word-dimension compression

---

1 **begin**
2   **for** *each topic z_j in $\mathcal{M}$* **do**
3 | calculate the total number of words s_{z_j} under the topic;
4   **end**
5   **for** *each word w_i in $\mathcal{M}$* **do**
6   calculate the sampling probability of the word according to Equation (9.3) and execute normalization;
7   **if** $a_{w_i}^{z_j} \leq \delta_A$ **then**
8 | delete the item $(z_j : p_{w_i}^{z_j})$ from w_i
9   **end**
10   **end**
11 **end**

We evaluate word-dimension compression through the task of text classification data set as well. As shown in Fig. 9.7, the experiment results show that with the increase of δ, the compression strength increases and the number of the remaining items decreases. It can be seen that when the threshold δ_A is set to 10^{-2}, 65.39% of the items can be removed, and the model, after compression, can still achieve similar performance as the original model.

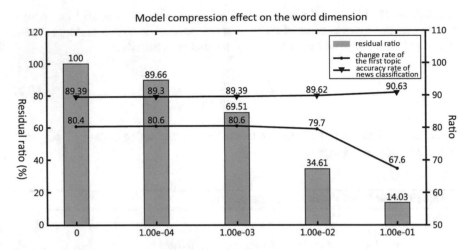

Fig. 9.7 Model compression on the word dimension

References

1. Charikar MS (2002) Similarity estimation techniques from rounding algorithms. In: Proceedings of the Thirty-Fourth Annual ACM Symposium on Theory of Computing. ACM, New York, pp 380–388
2. Chierichetti F, Kumar R, Pandey S, Vassilvitskii S (2010) Finding the Jaccard median. In: Proceedings of the Twenty-First Annual ACM-SIAM Symposium on Discrete Algorithms. SIAM, Philadelphia, pp 293–311
3. Song Y, Tong Y, Bao S, Jiang D, Wu H, Wong RCW (2020) TopicOcean: an ever-increasing topic model with meta-learning. In: 2020 IEEE International Conference on Data Mining (ICDM). IEEE, Piscataway, pp 1262–1267

Chapter 10
Applications

In the previous chapters, we discussed how to train high-quality topic models. How to utilize these models in real-life applications is also a challenging issue. This chapter discusses three paradigms of using topic models: semantic representation, semantic matching, and visualization.

10.1 Semantic Representation

As shown in Fig. 10.1, semantic representation refers to obtaining and applying the topic distribution of documents. The topic distribution of each document can be regarded as its semantic representation and is effective for various downstream machine learning algorithms. As the number of topics is usually fewer than the number of words in the vocabulary, the topic distribution is more concise than the one-hot representation of the document.

10.1.1 Text Classification

News feed has become the primary way to obtain news information in daily life. However, since news articles are obtained through various sources, high-quality and low-quality ones are usually mixed. To improve the user experience, news feed applications rely on text classifiers to distinguish between low-quality and high-quality news. Conventional text classifier depends on manually designed features, including the source site, content length, and the number of pictures. As manually creating features relies on extensive domain knowledge and expertise, their number is usually limited. The topic distribution can be cheaply obtained features for

Fig. 10.1 Semantic
representation

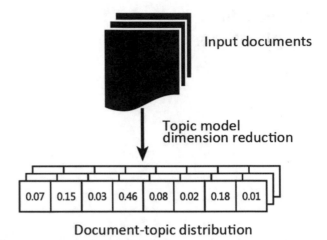

Input documents

Topic model
dimension reduction

| 0.07 | 0.15 | 0.03 | 0.46 | 0.08 | 0.02 | 0.18 | 0.01 |

Document-topic distribution

the document. Collectively utilizing the topic distribution and manually designed features improves the performance of the text classifier.

We use gradient boost decision tree (GBDT) [3] as an example to illustrate how to utilize the topic distribution. Mathematically, GBDT can be regarded as an additive model containing K trees:

$$y_i = \sum_{k=1}^{K} f_k(x_i), \; f_k \in F \tag{10.1}$$

where f_k represents the k-th tree, x_i represents the features of the i-th news, and y_i represents the quality of the i-th news.

We use manually labeled test data to evaluate the effect of introducing the topic distribution to x_i. The news quality is divided into three levels: 0, 1, and 2. Level 0 means the worst quality, while level 2 means the best quality. Figure 10.2 presents the accuracy of the GBDT model using different feature sets. The topic distribution improves the performance of the text classifier, suggesting that the topic distribution is an effective representation of the document.

10.1.2   Text Clustering

As news articles from different sources are highly redundant, we must avoid redundancy to improve the user experience. Semantically related news can be clustered together to facilitate reducing redundancy. Since the topic distribution of the document is low-dimensional and contains semantic information, it is an effective representation for text clustering. Based upon the K-means algorithm

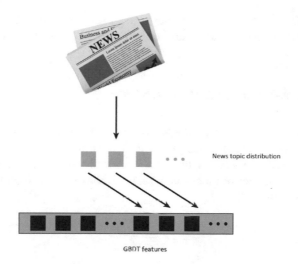

(a) Topic distribution as GBDT feature

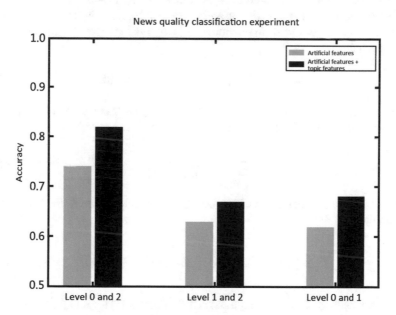

(b) Classification results

Fig. 10.2 Classification with regard to news quality

[4], we conduct text clustering through the topic distributions of documents. Experimental result shows that topic distribution is effective for text clustering.

10.1.3   Click-Through Rate Prediction

In web search or recommendation systems, we need to present the most valuable item[1] to the user. To achieve this, we typically use the click-through rate (CTR) prediction to determine whether the item is aligned with the user's interest. CTR is defined as the number of times that an item is clicked divided by the number of times that it is presented:

$$CTR = \frac{\#_{click}}{\#_{show}} \tag{10.2}$$

where $\#_{click}$ is the number of clicks of the item, and $\#_{show}$ is the number of times the item is presented. CTR prediction becomes a common task in web search and recommendation. CTR prediction usually relies on logistic regression (LR) model based on features derived from the items and empirical click data. Suppose that the empirical data are

$$(x_1, y_1), (x_2, y_2), \ldots, (x_n, y_n), y_i \in (0, 1) \tag{10.3}$$

where x_i represents the features of an item i, and y_i indicates whether the item i is clicked (1) or not (0). The CTR of the item i is calculated as follows:

$$CTR_{x_i} = \frac{1}{1 + e^{-\Omega^T x_i}} \tag{10.4}$$

where Ω is the parameter of the LR model. Cross entropy is used as the loss function to train the model:

$$\frac{1}{n} \sum_{i=1}^{n} (-y_i \log(CTR_{x_i}) - (1 - y_i) \log(1 - CTR_{x_i})) \tag{10.5}$$

The mechanism of introducing topic distribution for CTR prediction is presented in Fig. 10.3. We first use the topic model to obtain the topic distribution of an item j and then utilize the topic distribution of the item as additional features of the LR model. The topic distribution and the other features are then collectively trained in the LR model. A comparison of CTR prediction with and without topic

---

[1] Items refer to the things that need to be presented to the user, such as webpages, advertisements, commodities, etc.

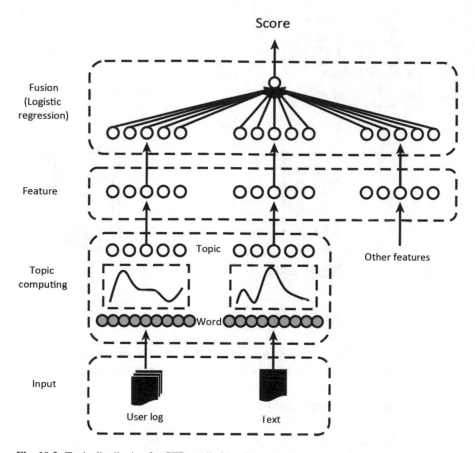

Fig. 10.3 Topic distribution for CTR prediction

features shows that the former is superior, suggesting that the topic information can effectively represent the items and the users' interests.

10.2 Semantic Matching

Many applications need to evaluate the semantic similarity of two texts, referred to as "semantic matching." According to the text length, semantic matching can be categorized into three types: short–short text matching, short–long text matching, and long–long text matching. Given that topic inference on short text is challenging, neural network models [6, 15] are more common in short–short text matching. In this section, we discuss how to apply topic models for short–long text matching and long–long text matching.

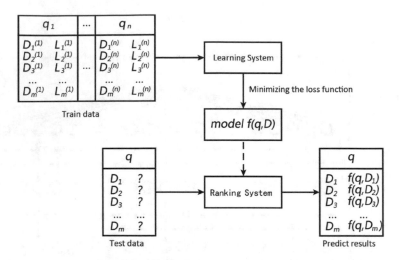

Fig. 10.4 LTR in advertising page ranking

10.2.1   Short–Long Text Matching

Short–long text matching is widely used in industry. We demonstrate its application with two examples: advertising page ranking and keyword extraction.

10.2.1.1   Advertising Page Ranking

In advertising page ranking, we rank advertising pages based on a learning-to-rank (LTR) model that depends on many features. The semantic similarity between the user query and the advertising page is one of the essential features of the LTR model. In Fig. 10.4, D represents the advertising page, q is the search query, and L represents the score calculated based upon D and q. We must train an LTR model $f(q, D)$ to predict the L between q and D. There are three types of features in an LTR model: those based on the query q, those based on the document D, and those based on $< q_i, D_j^{(i)} >$ pairs. The most important ones are those based on the document D and those based on $< q_i, D_j^{(i)} >$ pairs. We can introduce the semantic matching scores of the query and the document as additional features. Since topic inference for short texts is hard, the topic inference of q is avoided. Based on the topic distribution of D, the probability of the distribution generating q can be calculated as follows:

$$\text{Similarity}(q, D) = \prod_{w \in q} \sum_{k=1}^{K} P(w|z_k) P(z_k|D) \tag{10.6}$$

where w denotes a word in q.

During the prediction stage, the LTR scores are calculated to produce the final ranking result. Experiments show that the introduction of topic-based semantic matching features can effectively improve the performance of LTR models.

10.2.1.2 Keyword Extraction

Another application of short–long text semantic matching is keyword extraction from documents, where the short text refers to the word, while the long text refers to the document.

When extracting keywords from documents, the most common features are term frequency (TF) and inverse document frequency (IDF). TF refers to the frequency of a specified word in a document, and IDF is calculated as $IDF(w) = log\frac{N}{N_w+1}$, where N is the total number of documents in the corpus, and N_w is the number of documents containing the word w. By introducing the semantic importance of the word in a document as an additional feature, the following equation is used to evaluate the similarity between each word and the document:

$$\text{Similarity}(w, d) = \sum_{k=1}^{K} \cos(\mathbf{v_w}, \mathbf{z_k}) P(z_k|d) \tag{10.7}$$

where d represents the document, w represents the word, $\mathbf{v_w}$ means the embedding corresponding to w, and $\mathbf{z_k}$ represents the embedding corresponding to the topic z_k, which can be obtained by topic models such as TWE.

10.2.2 Long–Long Text Matching

After obtaining the topic distributions of long texts, we can calculate the distance between them and use it as an indicator of semantic similarity. Besides the aforementioned JS divergence and cosine similarity, Hellinger distance [13] can be used to measure the distance between two topic distributions P and Q:

$$\text{Hellinger\_Distance}(P, Q) = \frac{1}{\sqrt{2}}\sqrt{\sum_{k=1}^{K}(\sqrt{p_i} - \sqrt{q_i})^2} \tag{10.8}$$

To illustrate the relations of the three metrics, we visualize them in Fig. 10.5. As can be seen from the figure, there is a positive correlation between JS divergence and Hellinger distance, while a negative correlation between cosine similarity and JS divergence. Empirically, industrial applications are not sensitive to the choice of the three metrics, and practitioners may choose any one in their applications.

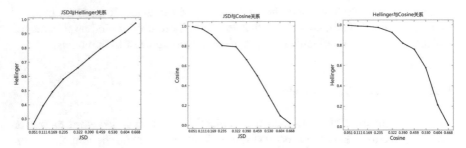

Fig. 10.5 Relation of long–long text matching metrics

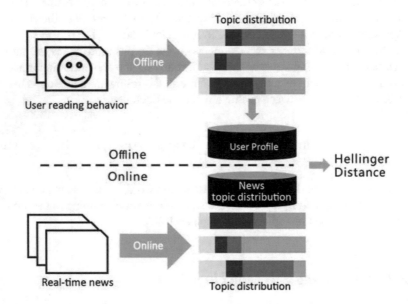

Fig. 10.6 Personalized recommendations

The ultimate goal of news feed applications is to recommend news that interests users. Hence, it is necessary to understand the information users are most interested in by analyzing their browsing history. It can be seen from Fig. 10.6 that when accumulating a certain amount of user behavior information (i.e., the news users recently clicked), we can produce an abstract "document" by combining the clicked news. Topic distribution of the abstract document can be used as the user profile that represents what the user is interested in. In real applications, by calculating the Hellinger distance between the topic distribution of each news in real time, we can determine the degree of users' potential interest in the news and provide personalized recommendations of news feeds.

10.3 Semantic Visualization

One major utility of topic models is to help people understand the content of a large corpus in an efficient manner, saving the time and effort required to read the details of each document. Semantic visualization reveals implicit semantics in a document through figures and charts. In this section, we discuss topic visualization tools and their applications.

10.3.1 Basic Visualization

Semantic visualization needs to illustrate the topic distribution of the document and the contents of each topic. Figure 10.7a displays the visualization of the topic distribution of a news article. The topic distribution is plotted in the form of a histogram, and users can intuitively see that the main contents of the news focus on the topics of "automobile," "artificial intelligence," and "Internet." Meanwhile, as shown in Fig. 10.7b and c, the word distribution under the topic of "automobile" and "artificial intelligence" can be presented as pie charts or word clouds, and users can view the contents of each topic in depth. Overall, this visualization scheme allows users to quickly understand the content of a corpus and the documents in it.

However, most topic models cannot annotate the topics with proper names, meaning that the document-topic visualization cannot be completely automated. To solve this problem, researchers have proposed several methods to name the topics automatically; for example, in [11], a set of metrics is established to evaluate the quality of topic naming—user-friendliness, semantic correlation, etc. In [9], researchers proposed a topic naming algorithm that generates topic names through Wikipedia and web search. Practitioners of topic model visualization can refer to the above literature to develop a suitable topic naming mechanism.

10.3.2 Advanced Visualization

Researchers have proposed some advanced visualization methods for topic models. Termite [1] displays different topics on the horizontal axis and presents the high-frequency words within the topic on the vertical axis. The circles indicate the degree of semantic relation between the topic and the word. For each topic, Termite can also display the distribution of the high-frequency words in the topic with a histogram. The advantage of Termite is the explicit depiction of the connection between words and topics, and the disadvantage is its inability to intuitively show the relation between topics and documents.

Topic Explorer [12] effectively visualizes the relationship between the topic and the document. It represents such relations through multiple windows. A ribbon is

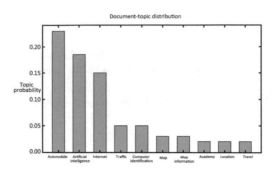

(a) Visualization of document-topic distribution

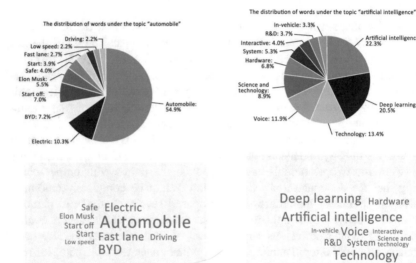

(b) The distribution of words under the topic "automobile"

(c) The distribution of words under the topic "artificial intelligence"

Fig. 10.7 Basic visualization

used to represent the topic distribution of each document, and the width of the ribbon indicates the weight of the topic. Clicking any document pops up a window displaying a list of documents that are topically similar to the clicked one.

For a corpus containing a large amount of documents, HierachicalTopics [2] can be applied. HierachicalTopics facilitates simultaneous analysis of multiple texts and supports the exploration of extensive collections of documents. HierachicalTopics depends on topic rose tree (TRT) algorithm to generate hierarchies for topics automatically.

10.3.3 General-Purpose Visualization Tools

There are many general-purpose visualization tools that can be used for visualizing topic model results. Figure 10.8 illustrates visualization tools with different emphases. The tools that specifically focus on the analysis usually require the users to have a certain level of coding ability, such as R and Matplotlib. In contrast, the tools that focus on visual display are relatively user-friendly, such as D3 and ECharts [10]. As shown in Fig. 10.9, analysis-emphasized tools tend to be more flexible and allow the users to implement customized analysis, while the visualization-emphasized tools have limited flexibility. Users should select the most appropriate visualization tools based on their needs and the application scenarios.

Due to the challenges caused by high-dimensional data, topic model visualization tools often rely on dimension reduction. T-SNE [5] is a method that visualizes the patterns of high-dimensional data and maintains its local structure. In practice, before executing the visualization, t-SNE is applied to reduce the data dimension to two dimension or three dimension. Figure 10.10 exhibits the results of applying t-

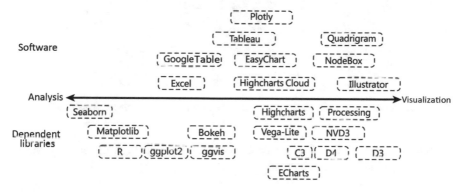

Fig. 10.8 Visualization tools in terms of analysis and visualization capability

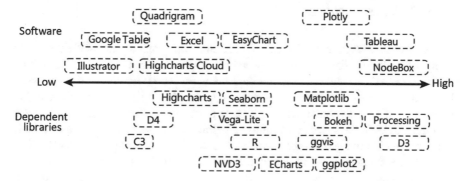

Fig. 10.9 Visualization tools in items of flexibility

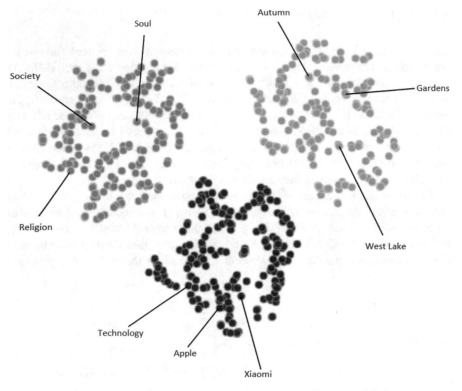

Fig. 10.10 Visualization of selected word embedding after t-SNE dimension reduction

SNE to word embedding. We can see that t-SNE retains the high-dimensional vector features after dimension reduction.

10.3.4   Other Applications

In addition to the aforementioned applications, topic models are widely used in many other fields. For example, researchers utilize topic models in information retrieval to evaluate the semantic relation between search queries and documents. Experiments show the superior performance of the topic representation effect relative to the traditional method [7, 8, 18]. In speech recognition, researchers introduce LDA to the speech recognition pipeline, and experiments indicate that this method can effectively reduce the word error rate of the recognition results [16, 17]. Moreover, in image processing, researchers use topic models to improve the effect of scene recognition [14]. Topic models are expected to play an increasingly significant role in different industries.

References

1. Chuang J, Manning CD, Heer J (2012) Termite: visualization techniques for assessing textual topic models. In: International Working Conference on Advanced Visual Interfaces, pp 74–77
2. Dou W, Yu L, Wang X, Ma Z, Ribarsky W (2013) HierarchicalTopics: visually exploring large text collections using topic hierarchies. IEEE Trans Vis Comput Graph 19(12):2002–2011
3. Friedman JH (2001) Greedy function approximation: a gradient boosting machine. Ann Stat, 1189–1232
4. Hartigan JA, Wong MA (1979) Algorithm as 136: a k-means clustering algorithm. J R Stat Soc Ser C (Appl Stat) 28(1):100–108
5. Hinton GE (2008) Visualizing high-dimensional data using t-SNE. J Mach Learn Res 9(2):2579–2605
6. Huang PS, He X, Gao J, Deng L, Acero A, Heck L (2013) Learning deep structured semantic models for web search using clickthrough data. In: Proceedings of the 22nd ACM International Conference on Information & Knowledge Management. ACM, New York, pp 2333–2338
7. Jiang D, Leung KWT, Ng W, Li H (2013) Beyond click graph: topic modeling for search engine query log analysis. In: Proceedings of the Database Systems for Advanced Applications: 18th International Conference, DASFAA 2013, Wuhan, April 22–25, 2013. Part I 18. Springer, Berlin, pp 209–223
8. Jiang D, Leung KWT, Yang L, Ng W (2015) TEII: topic enhanced inverted index for top-k document retrieval. Knowl Based Syst 89:346–358
9. Lau JH, Grieser K, Newman D, Baldwin T (2011) Automatic labelling of topic models. In: Proceedings of the 49th Annual Meeting of the Association for Computational Linguistics: Human Language Technologies. ACL, Toronto, pp 1536–1545
10. Li D, Mei H, Shen Y, Su S, Zhang W, Wang J, Zu M, Chen W (2018) ECharts: a declarative framework for rapid construction of web-based visualization. Visual Inform 2(2):136–146
11. Mei Q, Shen X, Zhai CX (2007) Automatic labeling of multinomial topic models. In: ACM SIGKDD International Conference on Knowledge Discovery and Data Mining, pp 490–499
12. Murdock J, Allen C (2015) Visualization techniques for topic model checking. In: AAAI Conference on Artificial Intelligence, pp 1–5
13. Nikulin MS (2001) Hellinger distance. Encycl Math 78
14. Niu Z, Hua G, Gao X, Tian Q (2012) Context aware topic model for scene recognition. In: 2012 IEEE Conference on Computer Vision and Pattern Recognition. IEEE, Piscataway, pp 2743–2750
15. Shen Y, He X, Gao J, Deng L, Mesnil G (2014) Learning semantic representations using convolutional neural networks for web search. In: Proceedings of the 23rd International Conference on World Wide Web. ACM, New York, pp 373–374
16. Song Y, Jiang D, Wu X, Xu Q, Wong RCW, Yang Q (2019) Topic-aware dialogue speech recognition with transfer learning. In: INTERSPEECH, pp 829–833
17. Wintrode J, Khudanpur S (2014) Combining local and broad topic context to improve term detection. In: 2014 IEEE Spoken Language Technology Workshop (SLT). IEEE, Piscataway, pp 442–447
18. Yi X, Allan J (2009) A comparative study of utilizing topic models for information retrieval. In: European Conference on Information Retrieval. Springer, Berlin, pp 29–41

Appendix A
Topic Models

A.1 Common Topic Models

See Table A.1.

Table A.1 Common topic models

Model	Year	Description
Latent Semantic Analysis (LSA) [20]	1990	Mining the latent semantics for a text corpus through SVD
Probabilistic latent semantic analysis (PLSA) [29]	1999	Mining the latent semantics for a text corpus from a probabilistic perspective, assuming that each document has a topic distribution
Latent Dirichlet Allocation (LDA) [5]	2003	Mining the latent semantics for a text corpus from a fully Bayesian perspective, using a Dirichlet prior to generate the topic distribution of a document
MG-LDA [88]	2008	Categorizing the latent topics into global and local ones
Rethinking LDA [90]	2009	Analyzing the effects of symmetric and asymmetric priors of topic models
MuTo [6]	2009	Topic modeling for the documents containing two languages
PCLSA [109]	2010	Discovering cross-lingual topics by introducing the translation vocabulary as a regularization item
Online LDA [28]	2010	Extending LDA to large data training based on variational Bayesian and online optimization
LDA-GA [69]	2013	Using genetic algorithm to determine the optimal configuration of LDA in different scenarios

© The Author(s), under exclusive license to Springer Nature Singapore Pte Ltd. 2023
D. Jiang et al., *Probabilistic Topic Models*,
https://doi.org/10.1007/978-981-99-2431-8

A.2   Topic Models with Advanced Features of Documents or Words

See Table A.2.

Table A.2   Topic models with advanced features of documents or words

Model	Year	Description
Author-Topic Model [77]	2004	Assuming that each author has a topic distribution
HMM-LDA [26]	2005	Introducing HMM for modeling the syntax structure
Statistical Entity-Topic Models [66]	2006	Mining the relation between topics and entities in documents
Bigram topic model [89]	2006	Introducing bigram into topic models
DCMLDA [22]	2009	Introducing the document-related topics to model burstiness phenomena in natural languages
Pitman–Yor topic model [78]	2010	Using the Pitman–Yor process to model the power-law distribution of words
SentenceLDA [36]	2011	Modeling word co-occurrence in sentences
DLDA [35]	2011	Jointly learning two groups of topics in short and long documents
Twitter-LDA [110]	2011	Extracting the topics from short text segments in Twitter
Factorial LDA [70]	2012	Introducing more features (e.g., authors' opinions and sentiment) for documents
BTM [13]	2014	Directly modeling the Biterm
Corr-wddCRF [47]	2016	Using the semantic distances between words to discover the semantically consistent disease topics
segLDAcop [1]	2017	Assuming that the generation process of a document is to first generate the segments and then the words in segments

A.3   Topic Models with Supervised Information

See Table A.3.

Table A.3 Topic models for introducing supervised information

Model	Year	Description
Opinion integration through semi-supervised topic modeling [55]	2008	Integrating perspectives using semi-supervised topic models
Supervised LDA [57]	2008	Introducing the supervised information corresponding to each document
DiscLDA [41]	2009	Modeling the supervised information by introducing the category labels on the document-topic distribution
Labeled LDA [73]	2009	Assuming that each document has several tags
WS-LDA [101]	2009	Training LDA for the disambiguation of named entity by supervised learning
Partially labeled topic models [74]	2011	Mining the latent topics with regard to each tag
TopicSpam [45]	2013	Detecting the subtle differences between deceptive and real texts
ELDA [76]	2014	Bridging the gap between social media material and reader sentiment and categorizing the sentiments of unlabeled documents
FLDA and DFLDA [46]	2015	Training the supervised topic model for multi-label document classification

A.4   Topic Models with Word Embedding

See Table A.4.

Table A.4 Topic models with word embedding

Model	Year	Description
LF-DMM [68]	2015	Assuming that each document has only one topic
Topical word embedding[53]	2015	Training the topic embedding by using sampled topic information with the word embedding
Latent topic embedding [33]	2016	Using a unified generation process to jointly train the topic model and the word embedding

A.5   Topic Models with Sentiment Information

See Table A.5.

Table A.5 Topic models with sentiment information

Model	Year	Description
Topic-sentiment mixture (TSM) [60]	2007	The topics are divided into background topic and content topic. The content topics are divided into three categories: neutral, positive, and negative
Multi-aspect sentiment model (MAS) [87]	2008	Modeling the relation between entity and sentiments
Joint sentiment/topic model [48]	2009	Mining sentiment information and topic information simultaneously in an unsupervised manner
Dependency sentiment LDA [44]	2010	Sentiment analysis based on the global topic information
ASUM [36]	2011	Introducing a prior of asymmetric sentiment information
FB-LDA/RCB-LDA [84]	2014	Analyzing and tracking the changes of public sentiment on Twitter
ADM-LDA [2]	2014	Improving the effect of sentiment analysis by using some sentence topics in the comments
Contextual sentiment topic model [75]	2016	Categorizing the topics into context-dependent sentiment topics and context-independent sentiment topics

A.6   Topic Models with Hierarchical Structure

See Table A.6.

Table A.6 Topic models with hierarchical structure

Model	Year	Description
Hierarchical LDA (hLDA) [4]	2003	The relation between topics is represented as a tree structure
Pachinko allocation model (PAM) [43]	2006	The relation between topics is represented as a directed acyclic graph
Hierarchical PAM [63]	2007	Combining the advantages of hLDA and PAM
Hierarchical latent tree analysis (HLTA) [52]	2014	Discovering hierarchical topic structure and solving the problem of topic being occupied by high-frequency words

A.7   Topic Models with Network Structure

See Table A.7.

Table A.7 Topic models with network structure

Model	Year	Description
Correlated topic model (CTM) [42]	2006	Modeling the correlation between each pair of topics
Topic evolution and social interactions [111]	2006	Discovering the dependencies between topics by shared documents in social networks
GWN-LDA [108]	2007	Community discovery on the data of social networks
Author Persona Topic Model [62]	2007	Assuming that each author can write under one or more roles
Author-Recipient-Topic Model [58]	2007	Modeling the senders and receivers of information for topic discovery based on social relationships
NetPLSA [61]	2008	Introducing a regularization item based on network structure
Pairwise-Link-LDA and Link-PLSA-LDA [65]	2008	Modeling the document and its reference information
Markov random topic fields [18]	2009	Using one or more user-specified diagrams to describe the relation between documents
Relational topic model [8]	2009	The relation between documents is represented by binary values
Networks uncovered by Bayesian inference (Nubbi) [9]	2009	Inferring the relationship between entities
Topic-Link LDA [50]	2009	Unifying topic modeling and author community discovery
iTopicModel [83]	2009	Modeling the relation between documents using Markov random field
Cite-LDA/Cite-PLSA-LDA [37]	2010	Modeling the reference relation between documents
MRF-LDA [100]	2015	Modeling the vocabulary knowledge and incorporating it into topic modeling
HawkesTopic Model [27]	2015	Introducing the information including "forward" and "follow" in social networks

A.8   Topic Models with Time Information

See Table A.8.

Table A.8 Topic models with time information

Model	Year	Description
Dynamic topic models (DTM) [3]	2006	Topic contents change over time
Topic over time (TOT) [92]	2006	Topic contents are fixed, but their intensity changes over time
Dynamic mixture models (DMMs) [97]	2007	Online topic discovery with multiple time series
Coordinated mixture model [94]	2007	Mining the burst patterns and the burst cycles in text streams
Continuous-time dynamic topic models (cDTM) [93]	2008	DTM with Brownian motion
Topic monitor [25]	2009	Topic adaptation in an evolving vocabulary

A.9   Topic Models with Geographic Information

See Table A.9.

Table A.9 Topic models with geographic information

Model	Year	Description
Spatiotemporal theme model [59]	2006	The topic distribution of a document is influenced by spatiotemporal factors
Geo-located image analysis using latent representations [16]	2008	Image analysis with geographic information
GeoFolk model [80]	2010	Using two-dimensional Gaussian distribution to represent the latitude and longitude information
Latent variable model for geographic lexical variation [24]	2010	Finding geographically consistent words in different language regions
Latent geographical topic analysis [103]	2011	Assuming that the topics are generated from "regions" based on documents with GPS coordinates
Geographic Web search topic discovery (G-WSTD) [32]	2012	Discovering geographically relevant topics for search logs
Who, Where, When, and What Model (W4) [104]	2013	Discovering the mobile behavior of individual users from three aspects: space, time, and activity

A.10   Topic Models with Bayesian Nonparametrics

See Table A.10.

Table A.10 Topic models with Bayesian nonparametrics

Model	Year	Description
Hierarchical LDA (hLDA) [4]	2003	The relation between topics is represented as a tree structure, and nested Chinese restaurant process is used as a prior
Syntactic topic model (STM) [7]	2009	Introducing parsing tree information
Indian buffet process compound Dirichlet process [99]	2010	Introducing a prior that combines hierarchical Dirichlet process and Indian buffet process
Nonparametric topic over time (npTOT) [23]	2013	Bayesian nonparametric version of the topic over time model
Enhanced W4 [106]	2015	Bayesian nonparametric version of W4 model

A.11   Distributed Training of Topic Models

See Table A.11.

Table A.11 Distributed training of topic models

Model	Year	Description
AD-LDA [67]	2009	Training LDA based on MapReduce
An Architecture for Parallel Topic Models [81]	2010	Using ParameterServer to train LDA
PLDA+ [51]	2011	Training LDA based on MPI
Mr.LDA [107]	2012	Training LDA based on variational inference and MapReduce
Spark-LDA [72]	2014	Training LDA based on Spark
LightLDA [105]	2015	Training LDA based on ParameterServer and Alias methods
WarpLDA [12]	2015	Training LDA with considering the features of hardware
Federated topic modeling [34]	2019	Training LDA in federated scenario

A.12   Visualization of Topic Models

See Table A.12.

Table A.12   Visualization of topic models

Model	Year	Description
Termite [15]	2012	Showing the relation between words and topics
UTOPIAN [14]	2013	Controlling the topic modeling results in a user-driven manner.
HierachicalTopics [21]	2013	Analyzing multiple types of texts efficiently in topic space
LDAvis [79]	2014	Interactively visualizing topics based on R and D3
Topic Explorer [64]	2015	Allowing users to generate explanatory assumptions and suggesting further experiments

A.13   Applications in Recommendation System

See Table A.13.

Table A.13   Applications in recommendation system

Model	Year	Description
Community-based dynamic recommendation [82]	2006	Using content semantics, evolutionary patterns, and user communities for recommendation
Google News Personalization [17]	2007	Generating personalized recommendation for the users of Google news
Combinational collaborative filtering [11]	2008	Executing the personalized community recommendation by simultaneously considering multiple types of co-occurrences in social data
LDA for tag recommendation [40]	2009	LDA-based recommendation for improving search
Collaborative topic modeling [91]	2011	Combining LDA and matrix decomposition
Dual role model [102]	2012	Distinguishing different roles of the users
DIGTOBI [39]	2013	Utilizing the voting mechanism for news website
TWILITE [38]	2014	Recommendation for Twitter
TOT-MMM [54]	2015	Recommendation of Twitter tag with time clustering

A.14   Applications in Information Retrieval and Information Extraction

See Table A.14.

Table A.14 Applications in information retrieval and information extraction

Model	Year	Description
LDA-based document models [96]	2006	Document model based on LDA for information retrieval
Query Latent Dirichlet Allocation (qLDA) [85]	2006	A topic model for generating abstracts using the query information
Bayesian query-focused summarization model [19]	2006	Using topic models to extract sentences and generate abstracts
SWB [10]	2007	Information retrieval based on the assumption that documents are a combination of multiple topics
TwitterRank [98]	2010	Using topic model to measure the influence of users
Bio-LDA [95]	2011	A topic model for extracting biological terms
Salience rank [86]	2017	A topic model for extracting keywords

A.15   Applications in Event Analysis

See Table A.15.

Table A.15 Applications in event analysis

Model	Year	Description
Popular events tracking (PET) [49]	2010	Modeling and tracking the popular events in social networks
Topic-perspective model [56]	2010	Modeling the generative process of social tags
ET-LDA [31]	2012	Modeling the topic information of Twitter and the corresponding events
Multi-modal Event Topic Model (mmETM) [71]	2016	Modeling multimedia documents with images and long text
PMB-LDA [30]	2016	Predicting the social and economic levels based on topic models

References

1. Amoualian H, Wei L, Gaussier E, Balikas G, Amini MR, Clausel M (2018) Topical coherence in LDA-based models through induced segmentation. In: Meeting of the Association for Computational Linguistics, pp 1799–1809
2. Bagheri A, Saraee M, De Jong F (2014) ADM-LDA: an aspect detection model based on topic modelling using the structure of review sentences. J Inf Sci 40(5):621–636
3. Blei DM, Lafferty JD (2006) Dynamic topic models. In: Proceedings of the 23rd International Conference on Machine Learning. ACM, New York, pp 113–120
4. Blei DM, Jordan MI, Griffiths TL, Tenenbaum JB (2003) Hierarchical topic models and the nested Chinese restaurant process. In: International Conference on Neural Information Processing Systems, pp 17–24
5. Blei DM, Ng AY, Jordan MI (2003) Latent Dirichlet allocation. J Mach Learn Res 3(Jan):993–1022
6. Boyd-Graber J, Blei DM (2009) Multilingual topic models for unaligned text. In: Proceedings of the Twenty-Fifth Conference on Uncertainty in Artificial Intelligence. AUAI Press, Quebec City, pp 75–82
7. Boyd-Graber JL, Blei DM (2009) Syntactic topic models. In: Advances in Neural Information Processing Systems, pp 185–192
8. Chang J, Blei DM (2009) Relational topic models for document networks. In: Artificial Intelligence and Statistics. JMLR.org, pp 81–88
9. Chang J, Boyd-Graber J, Blei DM (2009) Connections between the lines: augmenting social networks with text. In: Proceedings of the 15th ACM SIGKDD International Conference on Knowledge Discovery and Data Mining. ACM, New York, pp 169–178
10. Chemudugunta C, Smyth P, Steyvers M (2007) Modeling general and specific aspects of documents with a probabilistic topic model. In: Advances in Neural Information Processing Systems, pp 241–248
11. Chen WY, Zhang D, Chang EY (2008) Combinational collaborative filtering for personalized community recommendation. In: Proceedings of the 14th ACM SIGKDD International Conference on Knowledge Discovery and Data Mining. ACM, New York, pp 115–123
12. Chen J, Li K, Zhu J, Chen W (2015) WarpLDA: a cache efficient O(1) algorithm for latent Dirichlet allocation. Proc VLDB Endow 9(10):744–755
13. Cheng X, Yan X, Lan Y, Guo J (2014) BTM: topic modeling over short texts. IEEE Trans Knowl Data Eng (1):1–1
14. Choo J, Lee C, Reddy CK, Park H (2013) Utopian: user-driven topic modeling based on interactive nonnegative matrix factorization. IEEE Trans Vis Comput Graph 19(12):1992–2001
15. Chuang J, Manning CD, Heer J (2012) Termite: visualization techniques for assessing textual topic models. In: International Working Conference on Advanced Visual Interfaces, pp 74–77
16. Cristani M, Perina A, Castellani U, Murino V (2008) Geo-located image analysis using latent representations. In: IEEE Conference on Computer Vision and Pattern Recognition, 2008. CVPR 2008, pp 1–8
17. Das AS, Datar M, Garg A, Rajaram S (2007) Google news personalization: scalable online collaborative filtering. In: International World Wide Web Conferences, pp 271–280
18. Daumé III H (2009) Markov random topic fields. In: Proceedings of the ACL-IJCNLP 2009 Conference Short Papers, Association for Computational Linguistics, pp 293–296
19. Daumé H III, Marcu D (2006) Bayesian query-focused summarization. In: ACL, Association for Computational Linguistics, pp 305–312
20. Deerwester S, Dumais ST, Furnas GW, Landauer TK, Harshman R (1990) Indexing by latent semantic analysis. J Am Soc Inf Sci 41(6):391
21. Dou W, Yu L, Wang X, Ma Z, Ribarsky W (2013) Hierarchical topics: visually exploring large text collections using topic hierarchies. IEEE Trans Vis Comput Graph 19(12):2002–2011

22. Doyle G, Elkan C (2009) Accounting for burstiness in topic models. In: Proceedings of the 26th Annual International Conference on Machine Learning. ACM, New York, pp 281–288
23. Dubey A, Hefny A, Williamson S, Xing EP (2013) A nonparametric mixture model for topic modeling over time. In: Proceedings of the 2013 SIAM International Conference on Data Mining. SIAM, Philadelphia, pp 530–538
24. Eisenstein J, O'Connor B, Smith NA, Xing EP (2010) A latent variable model for geographic lexical variation. In: Conference on Empirical Methods in Natural Language Processing, pp 1277–1287
25. Gohr A, Hinneburg A, Schult R, Spiliopoulou M (2009) Topic evolution in a stream of documents. In: Proceedings of the 2009 SIAM International Conference on Data Mining. SIAM, Philadelphia, pp 859–870
26. Griffiths TL, Steyvers M, Blei DM, Tenenbaum JB (2005) Integrating topics and syntax. In: Advances in Neural Information Processing Systems, pp 537–544
27. He X, Rekatsinas T, Foulds J, Getoor L, Liu Y (2015) HawkesTopic: a joint model for network inference and topic modeling from text-based cascades. In: International Conference on Machine Learning, pp 871–880
28. Hoffman M, Bach FR, Blei DM (2010) Online learning for latent Dirichlet allocation. In: Advances in Neural Information Processing Systems, pp 856–864
29. Hofmann T (1999) Probabilistic latent semantic indexing. In: Proceedings of the 22nd Annual International ACM SIGIR Conference on Research and Development in Information Retrieval. ACM, New York, pp 50–57
30. Hong L, Frias-Martinez E, Frias-Martinez V (2016) Topic models to infer socio-economic maps. In: Thirtieth AAAI Conference on Artificial Intelligence, pp 3835–3841
31. Hu Y, John A, Wang F, Kambhampati S (2012) ET-LDA: joint topic modeling for aligning events and their Twitter feedback. In: Proceedings of the AAAI Conference on Artificial Intelligence, vol 12, pp 59–65
32. Jiang D, Vosecky J, Leung KWT, Ng W (2012) G-WSTD: a framework for geographic Web search topic discovery. In: Proceedings of the 21st ACM International Conference on Information and Knowledge Management, pp 1143–1152
33. Jiang D, Shi L, Lian R, Wu H (2016) Latent topic embedding. In: Proceedings of COLING 2016, the 26th International Conference on Computational Linguistics: Technical Papers, pp 2689–2698
34. Jiang D, Song Y, Tong Y, Wu X, Zhao W, Xu Q, Yang Q (2019) Federated topic modeling. In: Proceedings of the 28th ACM International Conference on Information and Knowledge Management, pp 1071–1080
35. Jin O, Liu NN, Zhao K, Yu Y, Yang Q (2011) Transferring topical knowledge from auxiliary long texts for short text clustering. In: Proceedings of the 20th ACM International Conference on Information and Knowledge Management. ACM, New York, pp 775–784
36. Jo Y, Oh AH (2011) Aspect and sentiment unification model for online review analysis. In: Proceedings of the Fourth ACM International Conference on Web Search and Data Mining. ACM, New York, pp 815–824
37. Kataria S, Mitra P, Bhatia S (2010) Utilizing context in generative bayesian models for linked corpus. In: AAAI'10: Proceedings of the Twenty-Fourth AAAI Conference on Artificial Intelligence, vol 10, p 1
38. Kim Y, Shim K (2014) Twilite: a recommendation system for Twitter using a probabilistic model based on latent Dirichlet allocation. Inf Syst 42:59–77
39. Kim Y, Park Y, Shim K (2013) DIGTOBI: a recommendation system for Digg articles using probabilistic modeling. In: WWW, International World Wide Web Conferences Steering Committee. ACM, New York, pp 691–702
40. Krestel R, Fankhauser P, Nejdl W (2009) Latent Dirichlet allocation for tag recommendation. In: Proceedings of the Third ACM Conference on Recommender Systems. ACM, New York, pp 61–68
41. Lacoste-Julien S, Sha F, Jordan MI (2009) DiscLDA: discriminative learning for dimensionality reduction and classification. In: Advances in Neural Information Processing Systems, pp 897–904

42. Lafferty JD, Blei DM (2006) Correlated topic models. In: Advances in Neural Information Processing Systems, pp 147–154
43. Li W, McCallum A (2006) Pachinko allocation: DAG-structured mixture models of topic correlations. In: Proceedings of the 23rd International Conference on Machine Learning. ACM, New York, pp 577–584
44. Li F, Huang M, Zhu X (2010) Sentiment analysis with global topics and local dependency. In: AAAI'10: Proceedings of the Twenty-Fourth AAAI Conference on Artificial Intelligence, vol 10, pp 1371–1376
45. Li J, Cardie C, Li S (2013) TopicSpam: a topic-model based approach for spam detection. In: Proceedings of the 51st Annual Meeting of the Association for Computational Linguistics (Volume 2: Short Papers), vol 2, pp 217–221
46. Li X, Ouyang J, Zhou X (2015) Supervised topic models for multi-label classification. Neurocomputing 149:811–819
47. Li C, Rana S, Phung D, Venkatesh S (2016) Hierarchical Bayesian nonparametric models for knowledge discovery from electronic medical records. Knowl Based Syst 99(C):168–182
48. Lin C, He Y (2009) Joint sentiment/topic model for sentiment analysis. In: Proceedings of the 18th ACM Conference on Information and Knowledge Management. ACM, New York, pp 375–384
49. Lin CX, Zhao B, Mei Q, Han J (2010) PET: a statistical model for popular events tracking in social communities. In: Proceedings of the 16th ACM SIGKDD International Conference on Knowledge Discovery and Data Mining. ACM, New York, pp 929–938
50. Liu Y, Niculescu-Mizil A, Gryc W (2009) Topic-link LDA: joint models of topic and author community. In: Proceedings of the 26th Annual International Conference on Machine Learning. ACM, New York, pp 665–672
51. Liu Z, Zhang Y, Chang EY, Sun M (2011) PLDA+: parallel latent Dirichlet allocation with data placement and pipeline processing. ACM Trans Intell Syst Technol 2(3):26
52. Liu T, Zhang NL, Chen P (2014) Hierarchical latent tree analysis for topic detection. In: Joint European Conference on Machine Learning and Knowledge Discovery in Databases. Springer, Berlin, pp 256–272
53. Liu Y, Liu Z, Chua TS, Sun M (2015) Topical word embeddings. In: Proceedings of the AAAI Conference on Artificial Intelligence, pp 2418–2424
54. Lu HM, Lee CH (2015) The topic-over-time mixed membership model (TOT-MMM): a Twitter hashtag recommendation model that accommodates for temporal clustering effects. IEEE Intell Syst 30(3):18–25
55. Lu Y, Zhai C (2008) Opinion integration through semi-supervised topic modeling. In: International World Wide Web Conferences, pp 121–130
56. Lu C, Hu X, Chen X, Park JR, He T, Li Z (2010) The topic-perspective model for social tagging systems. In: Proceedings of the 16th ACM SIGKDD International Conference on Knowledge Discovery and Data Mining, KDD '10. ACM, New York, pp 683–692. https://doi.org/10.1145/1835804.1835891, http://doi.acm.org/10.1145/1835804.1835891
57. Mcauliffe JD, Blei DM (2008) Supervised topic models. In: Advances in Neural Information Processing Systems, pp 121–128
58. McCallum A, Wang X, Corrada-Emmanuel A (2007) Topic and role discovery in social networks with experiments on Enron and academic email. J Artif Intell Res 30:249–272
59. Mei Q, Liu C, Su H, Zhai C (2006) A probabilistic approach to spatiotemporal theme pattern mining on weblogs. In: Proceedings of the 15th International Conference on World Wide Web. ACM, New York, pp 533–542
60. Mei Q, Ling X, Wondra M, Su H, Zhai C (2007) Topic sentiment mixture: modeling facets and opinions in weblogs. In: Proceedings of the 16th International Conference on World Wide Web. ACM, New York, pp 171–180
61. Mei Q, Cai D, Zhang D, Zhai C (2008) Topic modeling with network regularization. In: Proceedings of the 17th International Conference on World Wide Web. ACM, New York, pp 101–110

62. Mimno D, McCallum A (2007) Expertise modeling for matching papers with reviewers. In: Proceedings of the 13th ACM SIGKDD International Conference on Knowledge Discovery and Data Mining. ACM, New York, pp 500–509

63. Mimno D, Li W, McCallum A (2007) Mixtures of hierarchical topics with pachinko allocation. In: Proceedings of the 24th International Conference on Machine Learning. ACM, New York, pp 633–640

64. Murdock J, Allen C (2015) Visualization techniques for topic model checking. In: AAAI Conference on Artificial Intelligence, pp 1–5

65. Nallapati RM, Ahmed A, Xing EP, Cohen WW (2008) Joint latent topic models for text and citations. In: Proceedings of the 14th ACM SIGKDD International Conference on Knowledge Discovery and Data Mining. ACM, New York, pp 542–550

66. Newman D, Chemudugunta C, Smyth P (2006) Statistical entity-topic models. In: Proceedings of the 12th ACM SIGKDD International Conference on Knowledge Discovery and Data Mining. ACM, New York, pp 680–686

67. Newman D, Asuncion A, Smyth P, Welling M (2009) Distributed algorithms for topic models. J Mach Learn Res 10(Aug):1801–1828

68. Nguyen DQ, Billingsley R, Du L, Johnson M (2015) Improving topic models with latent feature word representations. Trans Assoc Comput Linguis 3:299–313

69. Panichella A, Dit B, Oliveto R, Di Penta M, Poshyvanyk D, De Lucia A (2013) How to effectively use topic models for software engineering tasks? An approach based on genetic algorithms. In: Proceedings of the 2013 International Conference on Software Engineering. IEEE Press, Piscataway, pp 522–531

70. Paul M, Dredze M (2012) Factorial LDA: sparse multi-dimensional text models. In: Advances in Neural Information Processing Systems, pp 2582–2590

71. Qian S, Zhang T, Xu C, Shao J (2016) Multi-modal event topic model for social event analysis. IEEE Trans Multimedia 18(2):233–246

72. Qiu Z, Wu B, Wang B, Shi C, Yu L (2014) Collapsed Gibbs sampling for latent Dirichlet allocation on spark. In: Proceedings of the 3rd International Conference on Big Data, Streams and Heterogeneous Source Mining: Algorithms, Systems, Programming Models and Applications, vol 36, pp 17–28

73. Ramage D, Hall D, Nallapati R, Manning CD (2009) Labeled LDA: a supervised topic model for credit attribution in multi-labeled corpora. In: Proceedings of the 2009 Conference on Empirical Methods in Natural Language Processing: Volume 1. Association for Computational Linguistics, Cedarville, pp 248–256

74. Ramage D, Manning CD, Dumais S (2011) Partially labeled topic models for interpretable text mining. In: Proceedings of the 17th ACM SIGKDD International Conference on Knowledge Discovery and Data Mining. ACM, New York, pp 457–465

75. Rao Y (2016) Contextual sentiment topic model for adaptive social emotion classification. IEEE Intell Syst 31(1):41–47

76. Rao Y, Lei J, Wenyin L, Li Q, Chen M (2014) Building emotional dictionary for sentiment analysis of online news. World Wide Web 17(4):723–742

77. Rosen-Zvi M, Griffiths T, Steyvers M, Smyth P (2004) The author-topic model for authors and documents. In: Proceedings of the 20th Conference on Uncertainty in Artificial Intelligence. AUAI Press, Quebec City, pp 487–494

78. Sato I, Nakagawa H (2010) Topic models with power-law using Pitman-Yor process. In: Proceedings of the 16th ACM SIGKDD International Conference on Knowledge Discovery and Data Mining. ACM, New York, pp 673–682

79. Sievert C, Shirley KE (2014) LDAvis: a method for visualizing and interpreting topics. In: The Workshop on Interactive Language Learning

80. Sizov S (2010) GeoFolk: Latent spatial semantics in Web 2.0 social media. In: WSDM '10: Proceedings of the Third ACM International Conference on Web Search and Data Mining. ACM, New York, pp 281–290

81. Smola A, Narayanamurthy S (2010) An architecture for parallel topic models. Proc VLDB Endow 3(1–2):703–710

82. Song X, Lin CY, Tseng BL, Sun MT (2006) Modeling evolutionary behaviors for community-based dynamic recommendation. In: Proceedings of the 2006 SIAM International Conference on Data Mining. SIAM, Philadelphia, pp 559–563

83. Sun Y, Han J, Gao J, Yu Y (2009) iTopicModel: information network-integrated topic modeling. In: 2009 Ninth IEEE International Conference on Data Mining. IEEE, Piscataway, pp 493–502

84. Tan S, Li Y, Sun H, Guan Z, Yan X, Bu J, Chen C, He X (2014) Interpreting the public sentiment variations on Twitter. IEEE Trans Knowl Data Eng 26(5):1158–1170

85. Tang J, Yao L, Chen D (2009) Multi-topic based query-oriented summarization. In: Proceedings of the 2009 SIAM International Conference on Data Mining. SIAM, Philadelphia, pp 1147–1158

86. Teneva N, Cheng W (2017) Salience rank: efficient keyphrase extraction with topic modeling. In: Meeting of the Association for Computational Linguistics, pp 530–535

87. Titov I, McDonald R (2008) A joint model of text and aspect ratings for sentiment summarization. In: Proceedings of ACL-08: HLT, pp 308–316

88. Titov I, McDonald R (2008) Modeling online reviews with multi-grain topic models. In: Proceedings of the 17th International Conference on World Wide Web. ACM, New York, pp 111–120

89. Wallach HM (2006) Topic modeling: beyond bag-of-words. In: Proceedings of the 23rd International Conference on Machine Learning. ACM, New York, pp 977–984

90. Wallach HM, Mimno DM, McCallum A (2009) Rethinking LDA: why priors matter. In: Advances in Neural Information Processing Systems, pp 1973–1981

91. Wang C, Blei DM (2011) Collaborative topic modeling for recommending scientific articles. In: Proceedings of the 17th ACM SIGKDD International Conference on Knowledge Discovery and Data Mining. ACM, New York, pp 448–456

92. Wang X, Mccallum A (2006) Topics over time: a non-Markov continuous-time model of topical trends. In: ACM SIGKDD International Conference on Knowledge Discovery and Data Mining, pp 424–433

93. Wang C, Blei DM, Heckerman D (2008) Continuous time dynamic topic models. In: Proceedings of the Twenty-Fourth Conference on Uncertainty in Artificial Intelligence (UAI2008). AUAI Press, Quebec City, pp 579–586

94. Wang X, Zhai C, Hu X, Sproat R (2007) Mining correlated bursty topic patterns from coordinated text streams. In: Proceedings of the 13th ACM SIGKDD International Conference on Knowledge Discovery and Data Mining. ACM, New York, pp 784–793

95. Wang H, Ding Y, Tang J, Dong X, He B, Qiu J, Wild DJ (2011) Finding complex biological relationships in recent PubMed articles using bio-LDA. PloS One 6(3):e17243

96. Wei X, Croft WB (2006) LDA-based document models for ad-hoc retrieval. In: Proceedings of the 29th Annual International ACM SIGIR Conference on Research and Development in Information Retrieval. ACM, New York, pp 178–185

97. Wei X, Sun J, Wang X (2007) Dynamic mixture models for multiple time-series. In: IJCAI'07: Proceedings of the 20th International Joint Conference on Artifical Intelligence, vol 7, pp 2909–2914

98. Weng J, Lim EP, Jiang J, He Q (2010) TwitterRank: finding topic-sensitive influential twitterers. In: Proceedings of the Third ACM International Conference on Web Search and Data Mining. ACM, New York, pp 261–270

99. Williamson S, Wang C, Heller KA, Blei DM (2010) The IBP compound Dirichlet process and its application to focused topic modeling. In: Proceedings of the 27th International Conference on Machine Learning (ICML-10). Citeseer, pp 1151–1158

100. Xie P, Yang D, Xing E (2015) Incorporating word correlation knowledge into topic modeling. In: Proceedings of the 2015 Conference of the North American Chapter of the Association for Computational Linguistics: Human Language Technologies, pp 725–734

101. Xu G, Yang SH, Li H (2009) Named entity mining from click-through data using weakly supervised latent Dirichlet allocation. In: Proceedings of the 15th ACM SIGKDD International Conference on Knowledge Discovery and Data Mining. ACM, New York, pp 1365–1374

102. Xu F, Ji Z, Wang B (2012) Dual role model for question recommendation in community question answering. In: SIGIR '12: Proceedings of the 35th International ACM SIGIR Conference on Research and Development in Information Retrieval. ACM, New York, pp 771–780

103. Yin Z, Cao L, Han J, Zhai C, Huang T (2011) Geographical topic discovery and comparison. In: WWW '11: Proceedings of the 20th International Conference on World Wide Web. ACM, New York, pp 247–256

104. Yuan Q, Cong G, Ma Z, Sun A, Thalmann NM (2013) Who, where, when and what: discover spatio-temporal topics for Twitter users. In: KDD '13: Proceedings of the 19th ACM SIGKDD International Conference on Knowledge Discovery and Data Mining. ACM, New York, pp 605–613

105. Yuan J, Gao F, Ho Q, Dai W, Wei J, Zheng X, Xing EP, Liu TY, Ma WY (2015) LightLDA: big topic models on modest computer clusters. In: Proceedings of the 24th International Conference on World Wide Web, International World Wide Web Conferences Steering Committee, pp 1351–1361

106. Yuan Q, Cong G, Zhao K, Ma Z, Sun A (2015) Who, where, when and what: a non-parametric Bayesian approach to context-aware recommendation and search for Twitter users. TOIS

107. Zhai K, Boyd-Graber J, Asadi N, Alkhouja ML (2012) Mr. LDA: a flexible large scale topic modeling package using variational inference in MapReduce. In: Proceedings of the 21st International Conference on World Wide Web. ACM, New York, pp 879–888

108. Zhang H, Giles CL, Foley HC, Yen J (2007) Probabilistic community discovery using hierarchical latent Gaussian mixture model. In: AAAI'07: Proceedings of the 22nd National Conference on Artificial Intelligence, vol 7, pp 663–668

109. Zhang D, Mei Q, Zhai C (2010) Cross-lingual latent topic extraction. In: Proceedings of the 48th Annual Meeting of the Association for Computational Linguistics, Association for Computational Linguistics, pp 1128–1137

110. Zhao WX, Jiang J, Weng J, He J, Lim EP, Yan H, Li X (2011) Comparing Twitter and traditional media using topic models. In: European Conference on Information Retrieval. Springer, Berlin, pp 338–349

111. Zhou D, Ji X, Zha H, Giles CL (2006) Topic evolution and social interactions: how authors effect research. In: Proceedings of the 15th ACM International Conference on Information and Knowledge Management. ACM, New York, pp 248–257

Printed in the United States
by Baker & Taylor Publisher Services